MATH WORKBOOK

Copyright © 2020 APEX Test Prep

All Rights Reserved

Prime and Composite Numbers

Determine whether each number below is a prime or composite number. Circle P if the number is a prime number or circle C if the number is a composite number.

1) 19 (P) C	2) 41 (P) C	3) 34 P (C)
4) 38 P (C)	5) 67 (P) C	6) 97 (P) C
7) 66 P (C)	8) 43 (P) C	9) 61 (P) C
10) 36 P (C)	11) 98 P (C)	12) 25 P (C)
13) 31 (P) C	14) 65 P (C)	15) 23 (P) C

Greatest Common Factor

Determine the greatest common factor for each pair of numbers.

1. 69 and 36 _____
2. 89 and 49 _____
3. 45 and 60 _____
4. 40 and 8 _____
5. 24 and 18 _____
6. 30 and 4 _____
7. 12 and 20 _____
8. 24 and 120 _____
9. 5 and 92 _____
10. 70 and 28 _____
11. 18 and 54 _____
12. 14 and 35 _____
13. 36 and 54 _____
14. 29 and 87 _____
15. 18 and 42 _____

Least Common Multiple

Determine the least common multiple for each pair of numbers.

1) 2 and 15
_____

2) 30 and 12
_____

3) 20 and 60
_____

4) 15 and 120
_____

5) 10 and 4
_____

6) 30 and 15
_____

7) 24 and 60
_____

8) 24 and 4
_____

9) 8 and 30
_____

10) 10 and 15
_____

11) 40 and 30
_____

12) 121 and 11
_____

13) 9 and 6
_____

14) 13 and 7
_____

15) 15 and 12
_____

Prime Factorization

Fill in the prime factorization trees where displayed or list the prime factorization for each number if just a number is given.

1) 115 = _____

2) 136 = _____

3) 45 = _____

4) 104 = _____

5) 28 = _____

6) 60 = _____

7) 114 = _____

8) 88 = _____

9) 98 = _____

10) 96 = _____

11) 81 = _____

12) 110 = _____

13) 63 = _____

14) 130 = _____

15) 42 = _____

Order of Operations

Evaluate the following problems using the order of operations

1) $(10 - 3) \times (9 - 6) + 7^2$
   $7 \times 3 + 49$
   $21 + 49 = \boxed{70}$ → 70

2) $(15 - 7) \times (12 - 6) + 6^2$

3) $(10 - 3)^2 + (12 - 15 \div 5)$
   $12 - 3$

4) $2 \times (6 \times 3 - 8^2) + 22$

5) $(9 + 56 - 5) \div 2 + 3^2$

6) $(11 + 53 - 4^2) \div (9 + 7)$

7) $(14 + 19 - 3^2) \div (12 \div 2)$

8) $(3^2 - 4)^2 + (16 + 20 \div 10)$

9) $3(8 \div 2^2)^3 + (5 \times 8)$

10) $(2 \times 6^2 + 8) \div (8 - 20 \div 5)^2$

11) $4 + (3 \times 2)^2 \div 4$

12) $2 \times (6 + 3) \div (2 + 1)^2$

13) $2^2 \times (3 - 1) \div 2 + 3$

14) $(12 + 3) \times (8 - 2) - 5^2$

15) $(19 - 8) \times (13 - 3) + 2^2$

16) $(2 + 4)^2 + (9 + 12 \div 4)$

17) $3 \times (13 \times 3 + 8^2) - 12$

18) $[(4 + 3)^2 + 1] + 2^3 - 5$

19) $[6^2 + (20 \div 5 + 4^2)] \div 7$

20) $(15 \div 5)^2 - [(12 + 2) + 3^2]$

Operations with Fractions

Complete the following problems involving computations with fractions.

1) $\frac{6}{23} + \frac{3}{4}$

2) $\frac{5}{14} + \frac{11}{21}$

3) $\frac{1}{4} + \frac{18}{42}$

4) $\frac{9}{10} + \frac{18}{35}$

5) $\frac{18}{27} + \frac{4}{6}$

6) $\frac{1}{2} + \frac{2}{3} + \frac{3}{4}$

7) $\frac{4}{10} + \frac{3}{4} + \frac{1}{2}$

8) $\frac{1}{3} + \frac{5}{10} + \frac{3}{5}$

9) $\frac{1}{2} + \frac{2}{3} + \frac{6}{10}$

10) $\frac{1}{3} + \frac{8}{10} + \frac{1}{4}$

11) $\frac{4}{8} + \frac{1}{3} + \frac{6}{16}$

12) $\frac{1}{3} + \frac{1}{4} + \frac{3}{5}$

13) $\frac{3}{23} + \frac{3}{4} + \frac{2}{4}$

14) $\frac{10}{18} + \frac{2}{4} + \frac{7}{12}$

15) $\frac{1}{2} + \frac{4}{5} + \frac{2}{4}$

16) $\frac{14}{8} - \frac{3}{6}$

17) $\frac{4}{5} - \frac{8}{12}$

18) $\frac{1}{4} - \frac{2}{14}$

19) $\frac{11}{13} - \frac{15}{26}$

20) $\frac{3}{5} - \frac{8}{15}$

21) $\frac{16}{24} - \frac{1}{6}$

22) $\frac{10}{12} - \frac{6}{9}$

23) $\frac{2}{5} - \frac{3}{40}$

24) $\frac{6}{55} - \frac{1}{11}$

25) $\frac{3}{4} - \frac{4}{15}$

26) $\frac{4}{5} - \frac{1}{4} - \frac{1}{10}$

27) $\frac{8}{10} - \frac{1}{4} - \frac{1}{5}$

28) $\frac{4}{5} - \frac{1}{3} - \frac{1}{10}$

29) $\frac{9}{10} - \frac{1}{3} - \frac{1}{10}$

30) $\frac{4}{5} - \frac{1}{2} - \frac{1}{5}$

Operations with Fractions

Complete the following problems involving computations with fractions.

1. $\dfrac{5}{8} \times \dfrac{3}{5}$ _____

2. $\dfrac{7}{10} \times \dfrac{3}{8}$ _____

3. $\dfrac{2}{3} \times \dfrac{3}{8}$ _____

4. $\dfrac{2}{15} \times \dfrac{6}{7}$ _____

5. $\dfrac{3}{11} \times \dfrac{5}{12}$ _____

6. $\dfrac{2}{3} \times \dfrac{1}{3}$ _____

7. $\dfrac{3}{5} \times \dfrac{1}{2}$ _____

8. $\dfrac{4}{5} \times \dfrac{4}{7}$ _____

9. $\dfrac{1}{2} \times \dfrac{9}{10}$ _____

10. $\dfrac{7}{9} \times \dfrac{1}{2}$ _____

11. $\dfrac{9}{10} \times \dfrac{3}{14}$ _____

12. $\dfrac{8}{15} \times \dfrac{1}{5}$ _____

13. $\dfrac{6}{7} \times \dfrac{8}{81}$ _____

14. $\dfrac{1}{20} \times \dfrac{7}{25}$ _____

15. $\dfrac{1}{5} \times \dfrac{5}{6}$ _____

16. $\dfrac{10}{12} \times \dfrac{6}{7}$ _____

17. $\dfrac{8}{10} \times \dfrac{1}{6}$ _____

18. $\dfrac{3}{12} \times \dfrac{5}{14}$ _____

19. $\dfrac{1}{2} \times \dfrac{2}{15}$ _____

20. $\dfrac{3}{4} \times \dfrac{17}{20}$ _____

21. $\dfrac{8}{20} \times \dfrac{5}{8}$ _____

22. $\dfrac{4}{5} \times \dfrac{9}{12}$ _____

23. $\dfrac{17}{18} \times \dfrac{4}{10}$ _____

24. $\dfrac{2}{7} \times \dfrac{9}{15}$ _____

25. $\dfrac{4}{8} \times \dfrac{13}{18}$ _____

26. $\dfrac{4}{6} \div \dfrac{6}{9}$ _____

27. $\dfrac{2}{5} \div \dfrac{1}{6}$ _____

28. $\dfrac{6}{14} \div \dfrac{2}{10}$ _____

29. $\dfrac{2}{7} \div \dfrac{6}{10}$ _____

30. $\dfrac{2}{9} \div \dfrac{9}{10}$ _____

Operations with Fractions

Complete the following problems involving computations with fractions.

1) $\dfrac{1}{9} \div \dfrac{4}{9}$

2) $\dfrac{2}{3} \div \dfrac{5}{7}$

3) $\dfrac{10}{11} \div \dfrac{3}{7}$

4) $\dfrac{1}{6} \div \dfrac{3}{5}$

5) $\dfrac{1}{2} \div \dfrac{2}{5}$

6) $\dfrac{1}{11} \div \dfrac{3}{7}$

7) $\dfrac{1}{6} \div \dfrac{7}{12}$

8) $\dfrac{2}{9} \div \dfrac{4}{7}$

9) $\dfrac{7}{11} \div \dfrac{1}{4}$

10) $5 \div \dfrac{7}{10}$

11) $\dfrac{1}{4} \div 9$

12) $10 \div \dfrac{2}{3}$

13) $\dfrac{1}{5} \div 9$

14) $\dfrac{8}{10} \div 3$

15) $3 \div \dfrac{1}{5}$

16) $3\dfrac{5}{9} \div 3\dfrac{3}{7}$

17) $2\dfrac{1}{3} \div 2\dfrac{1}{4}$

18) $3\dfrac{5}{6} \div 4\dfrac{2}{5}$

19) $2\dfrac{1}{9} \div 3\dfrac{1}{7}$

20) $2\dfrac{1}{8} \div 2\dfrac{3}{5}$

21) $3\dfrac{1}{3} \div 3\dfrac{3}{4}$

22) $2\dfrac{1}{2} \div 2\dfrac{3}{7}$

23) $2\dfrac{7}{9} \div 3\dfrac{3}{7}$

24) $3\dfrac{1}{2} \div 4\dfrac{1}{3}$

25) $2\dfrac{5}{8} \div 2\dfrac{5}{7}$

26) $2\dfrac{1}{2} \div 2\dfrac{3}{7}$

27) $3\dfrac{5}{8} \div 3\dfrac{1}{6}$

28) $4\dfrac{2}{5} \div 2\dfrac{6}{7}$

29) $3\dfrac{2}{7} \div 4\dfrac{3}{5}$

30) $4\dfrac{1}{2} \div 3\dfrac{2}{3}$

Operations with Decimals

Complete the following problems involving computations with decimals.

1) 62.289 + 33.259 = _____

2) 86.927 + 14.969 = _____

3) 40.847 + 47.716 = _____

4) 95.689 + 67.132 = _____

5) 39.323 + 76.324 = _____

6) 53.379 + 87.957 = _____

7) 65.955 + 86.635 = _____

8) 54.557 + 97.499 + 75.548 = _____

9) 13.145 + 38.778 + 85.332 = _____

10) 80.433 + 13.737 + 88.967 = _____

11) 74.197 + 79.562 + 79.967 = _____

12) 41.675 + 62.566 + 67.483 = _____

13) 36.824 + 88.524 + 66.925 = _____

14) 45.152 + 90.349 + 99.487 = _____

15) 31.238 + 32.123 + 98.857 = _____

16) 991 − 63.609 = _____

17) 390 − 34.06 = _____

18) 4335 − 9.448 = _____

19) 8592 − 4714.4216 = _____

20) 8483 − 53.28 = _____

21) 9359 − 284.9 = _____

22) 4706 − 5.4222 = _____

23) 8352 − 4891.3918 = _____

24) 8079 − 5.1 = _____

25) 981 − 441.949 = _____

26) 755 − 0.1448 = _____

27) 7914 − 2174.557 = _____

28) 2978 − 61.257 = _____

29) 5954 − 3826.56 = _____

30) 660 − 359.2 = _____

Operations with Decimals

Complete the following problems involving computations with decimals.

1) 441 − 57.866 = ___	2) 37.79 + 13.6 = ___	3) 66.66 x 28.27 = ___
4) 12.32 x 68.18 = ___	5) 43.25 x 14.43 = ___	6) 33.79 x 28.59 = ___
7) 75.58 x 25.87 = ___	8) 44.85 x 30.58 = ___	9) 24.43 x 74.27 = ___
10) 77.21 x 79.19 = ___	11) 58.28 x 68.37 = ___	12) 79.34 x 46.65 = ___
13) 47.87 x 56.24 = ___	14) 81.12 x 37.83 = ___	15) 91.26 x 98.37 = ___
16) 66.86 x 55.76 = ___	17) 4193.73 ÷ 9 = ___	18) 547.04 ÷ 8 = ___
19) 309.26 ÷ 94 = ___	20) 594.72 ÷ 6 = ___	21) 758.08 ÷ 8 = ___
22) 5744.96 ÷ 32 = ___	23) 1514.72 ÷ 8 = ___	24) 95.68 ÷ 13 = ___
25) 276.54 ÷ 33 = ___	26) 4785.04 ÷ 52 = ___	27) 2463.84 ÷ 48 = ___
28) 17688.58 ÷ 26 = ___	29) 7043.52 ÷ 46 = ___	30) 10349.94 ÷ 51 = ___

Converting between Fractions and Decimals

Convert the following fractions into their equivalent decimal or decimal to its equivalent fraction. Be sure to reduce all fractions to lowest terms.

1) 0.083 = _____

2) $\dfrac{3}{5}$ = _____

3) $\dfrac{3}{4}$ = _____

4) $\dfrac{4}{6}$ = _____

5) $\dfrac{5}{6}$ = _____

6) $\dfrac{3}{8}$ = _____

7) $\dfrac{1}{11}$ = _____

8) $\dfrac{7}{8}$ = _____

9) $\dfrac{4}{9}$ = _____

10) $\dfrac{17}{18}$ = _____

11) $\dfrac{10}{13}$ = _____

12) $\dfrac{12}{21}$ = _____

13) $\dfrac{1}{19}$ = _____

14) $\dfrac{9}{25}$ = _____

15) $\dfrac{2}{7}$ = _____

16) 0.3 = _____

17) 0.54 = _____

18) 0.2 = _____

19) 0.44 = _____

20) 0.01 = _____

21) 0.98 = _____

22) 0.64 = _____

23) 0.37 = _____

24) 0.125 = _____

25) 1.0 = _____

Percents

Complete the following problems involving percentages.
Round your answers to two decimal places.

1) What is 16% of 15?

= _____

2) What is 75 percent of 60?

= _____

3) What is 2% of 26?

= _____

4) What percent of 77 is 52?

= _____

5) 13 is 81% of what number?

= _____

6) What percent of 83 is 49?

= _____

7) 76 is 86 percent of what number?

= _____

8) 88 is 85% of what number?

= _____

9) 76 is 34% of what number?

= _____

10) 13 is 81 percent of what number?

= _____

Percents

Complete the following problems involving percentages.
Round your answers to two decimal places.

11 What is 45% of 44?

= _____

12 What is 17% of 24?

= _____

13 60 is 81% of what number?

= _____

14 62 is 2% of what number?

= _____

15 What is 13% of 78?

= _____

16 70 is 93% of what number?

= _____

17 What percent of 79 is 36?

= _____

18 What is 71% of 17?

= _____

19 What percent of 98 is 16?

= _____

20 94 is 78% of what number?

= _____

Ratios

1) 12:15 = 40: = :35 = :5 = :25 = :40

2) 21:27 = :63 = 70: = 35: = 56: = :9

3) 40:48 = :30 = :42 = 15: = :6 = 50:

4) 5:9 = :63 = :81 = :90 = 10: = 40:

5) 16:24 = 2: = 6: = :30 = :15 = :21

6) 30:40 = 21: = 9: = 35: = :8 = 24:

7) 21:30 = :80 = 14: = :50 = :70 = :100

8) 6:42 = :70 = 7: = :21 = 4: = :56

9) 35:49 = :21 = 50: = :14 = 25: = 40:

10) 7:28 = 3: = :80 = 5: = :32 = 9:

11) 24:56 = 9: = 27: = :140 = :49 = 3:

12) 4:36 = :72 = 3: = :81 = :18 = 4:

13) 27:45 = :40 = 21: = :100 = :5 = :20

14) 4:14 = 10: = :7 = 8: = 16: = :21

15) 30:48 = :72 = :8 = 15: = 10: = :56

Rates

Determine the unit rate for each of the following.

1) $9.00 for 4 cans of soup
= _____

2) 155 miles driven on 6 gallons of gas
= _____

3) 11 pages read in 9 minutes
= _____

4) $16.00 for 7 pens
= _____

5) 14 math problems completed in 9 minutes
= _____

6) 100 miles driven on 6 gallons of gas
= _____

7) 8 inches of snow in 5 hours
= _____

8) 13 cupcakes cost 20 dollars
= _____

9) 15 cans of tuna cost $34.50
= _____

10) 13 pushups in 17 seconds
= _____

11) $145 for 4 chairs
= _____

12) 7 slices of pizza for $15.00
= _____

13) $13.50 for 5 packs of gum
= _____

14) $19.00 for 6 headbands
= _____

15) $140.00 for 4 textbooks
= _____

Proportions

Solve the following word problems related to proportional rates.

1) A plane flies 490 miles in 2 hours. At this rate, how far would the plane fly in 16 hours?

2) 5 cans of corn costs $2.80 at the corner bodega. 9 cans of the same corn costs $5.70 at the large grocery store. Which shop has the better value?

3) A roller coaster at the theme park can take 32 riders every 8 minutes. How many riders can the roller coaster accommodate in four hours?

4) The local bakery can make 320 cupcakes in 5 hours. How many can they bake in 24 hours?

5) Peter can read 14 pages in 25 minutes. To the nearest minute, how long will it take him to read a 175-page book? Report your answer in hours and minutes.

Proportions

Solve the following word problems related to proportional rates.

6) Every 30 minutes, 70 people can check out at the self-service check out at the grocery store. How many people can the self-service accommodate in the 9.5 hours the store is open?

7) Three lollipops of Sweeties' brand cost $1.23. 5 lollipops of Darla's brand cost $2.15. Which brand is a better value?

8) An earring factory makes 290 pairs of earrings in 6 hours. How many pairs can they make in the 40-hour work week?

9) Tanya does 146 jumping jacks in 3 minutes. How many does she do every 20 seconds?

10) Xiaoping earns 282 points in her computer game for every three levels she completes. How many points will she accrue after beating 50 levels?

Exponents

Evaluate the exponents and simplify where necessary.

1) 7^2
= _____

2) 4^3
= _____

3) $-(5)^3$
= _____

4) 3^{-3}
= _____

5) $(-2)^{-3}$
= _____

6) $-(8)^3$
= _____

7) 2^{-7}
= _____

8) 7^{-3}
= _____

9) $-(2)^8$
= _____

10) $7^{-3} \times 7^2$
= _____

11) $\dfrac{2^5}{2}$
= _____

12) $\dfrac{4^{-2}}{4}$
= _____

13) $6x^2 \times 3x$
= _____

14) $a^2 \times a^5 \times a^3$
= _____

15) $b \times b^{-4}$
= _____

Scientific Notation

Convert the numbers in scientific notation to standard format and those in standard format to scientific notation.

1) 3.451×10^2
= 345.1

2) 8.10×10^4
= 81,000

3) 9.25×10^1
= 92.5

4) 7.089×10^3
= 7,089

5) 3.001×10^5
= 300,100

6) 8.882×10^7
= 88,820,000

7) 4.02834×10^6
= 4,028,340

8) 1.90001×10^9
= 1,900,010,000

9) 6.107×10^1
= 61.07

10) 2.87×10^2
= 287

11) 785
= 7.85×10^2

12) 16,086
= 1.6086×10^4

13) 1.098
= 1.098×10^0

14) 0.00478
= 4.78×10^{-3}

15) 9,005,983
= 9.005983×10^6

16) 0.00066
= 6.6×10^{-4}

17) 10
= 1×10^1

18) 7,630,002
= 7.630002×10^6

19) 0.07
= 7×10^{-2}

20) 540
= 5.4×10^2

Absolute Value

Solve the following equations. Where necessary, round your answers to two decimal places.

1) $|9a| - 25 = 5$

2) $|27 - 2x| = 23$

3) $|-7y| = 28$

4) $\dfrac{4}{5}|b| = 9$

5) $\left|-\dfrac{c}{2} + 16\right| = 7$

6) $2|-z| = 8$

7) $3|k - 25| = 4$

8) $\left|\dfrac{v-5}{7}\right| = 10$

Absolute Value

Solve the following equations. Where necessary, round your answers to two decimal places.

9) $\dfrac{1}{|z - 18|} = 5$

10) $\dfrac{3}{|12 + a|} = 18$

11) $\left|\dfrac{-b}{6} - 12\right| = 10$

12) $\left|\dfrac{p}{7} + 28\right| = 6$

13) $|12 + 6s| = 26$

14) $\left|\dfrac{m}{7} - 7\right| = 29$

15) $\left|\dfrac{-x}{4} + 9\right| = 3$

Plotting Rational Numbers on the Coordinate Plane

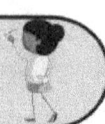

Circle the ordered pair that correctly locates the point plotted on the coordinate plane.

1) (1, 5)   (-1, 5)   (1, -5)   (-1, -5)

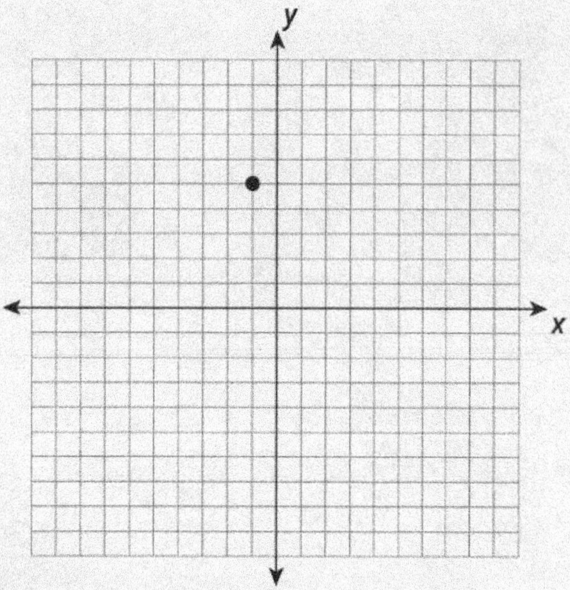

2) (3, 4)   (-3, 4)   (3, -4)   (-3, -4)

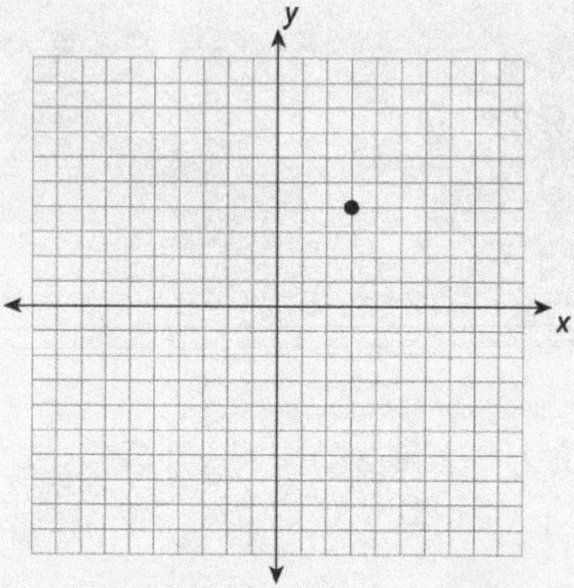

3) (0, 6)   (6, 0)   (0, -6)   (-6, 0)

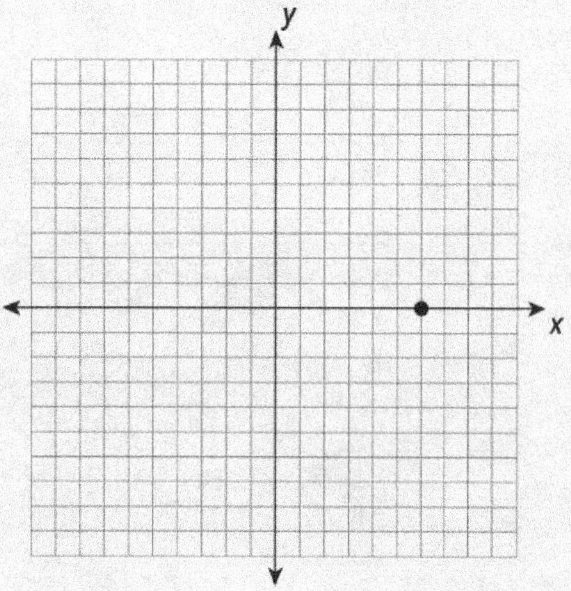

4) (4, 1)   (4, -1)   (-4, 1)   (-4, -1)

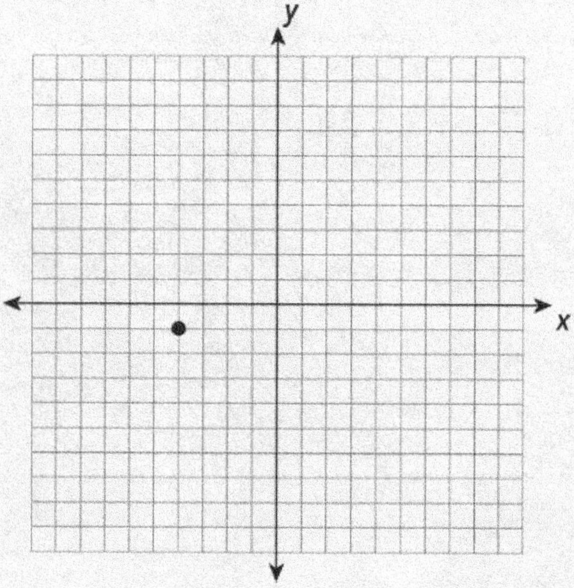

Plotting Rational Numbers on the Coordinate Plane

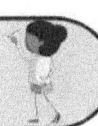

Circle the ordered pair that correctly locates the point plotted on the coordinate plane.

5) (6, 3) (6, -3) (-6, 3) (-6, -3)

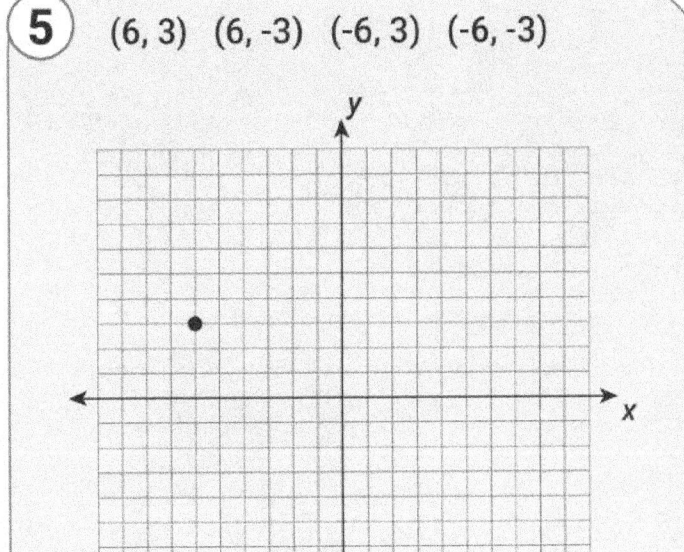

6) (2, 7) (2, -7) (-2, 7) (-2, -7)

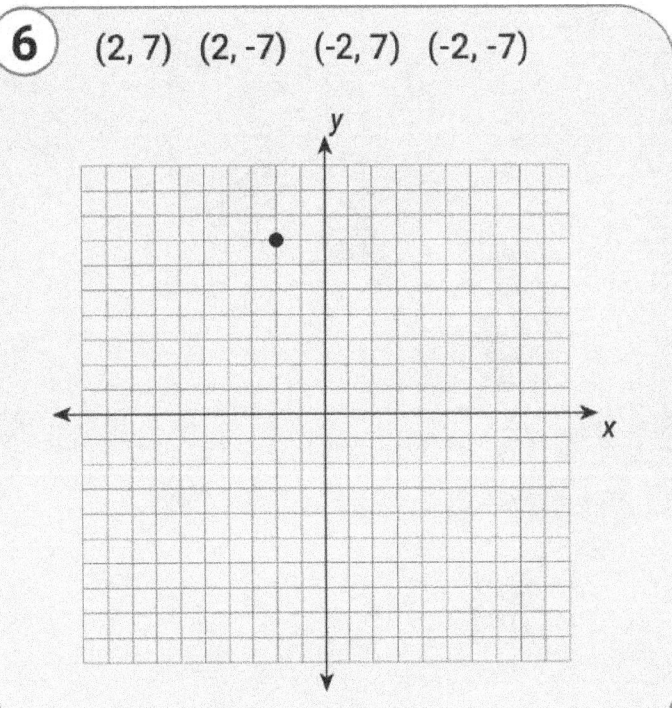

7) (2, 3) (2, -3) (-2, 3) (-2, -3)

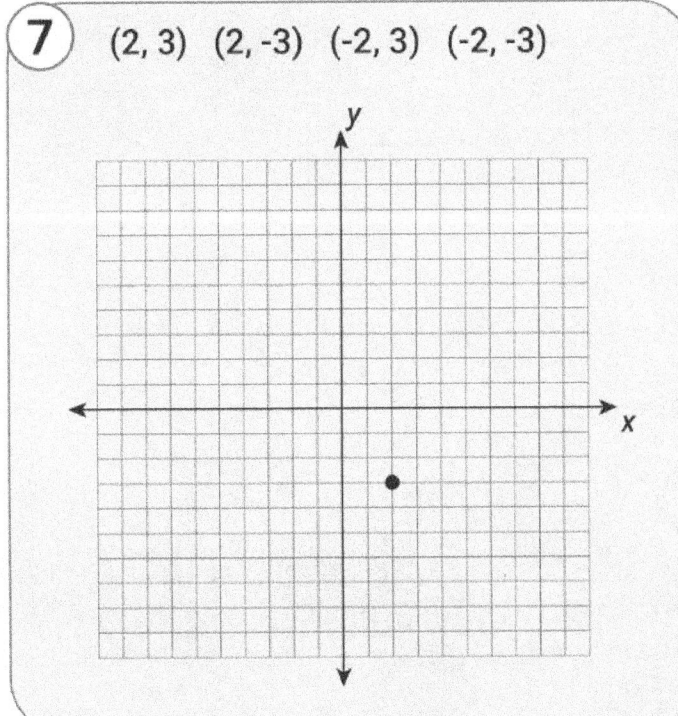

8) (9, 5) (-9, 5) (9, -5) (-9, -5)

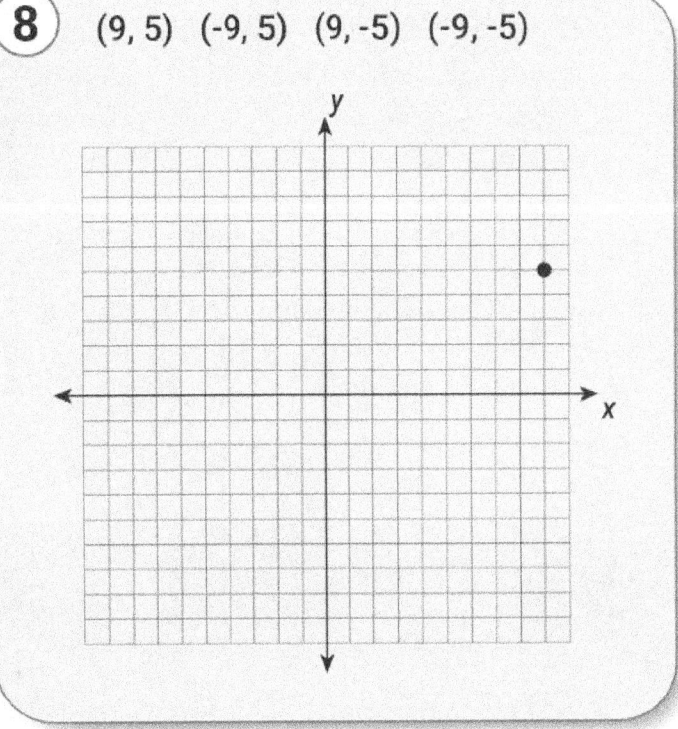

Plot the Point

Plot the point on the provided coordinate plane.

1) (2, 4)

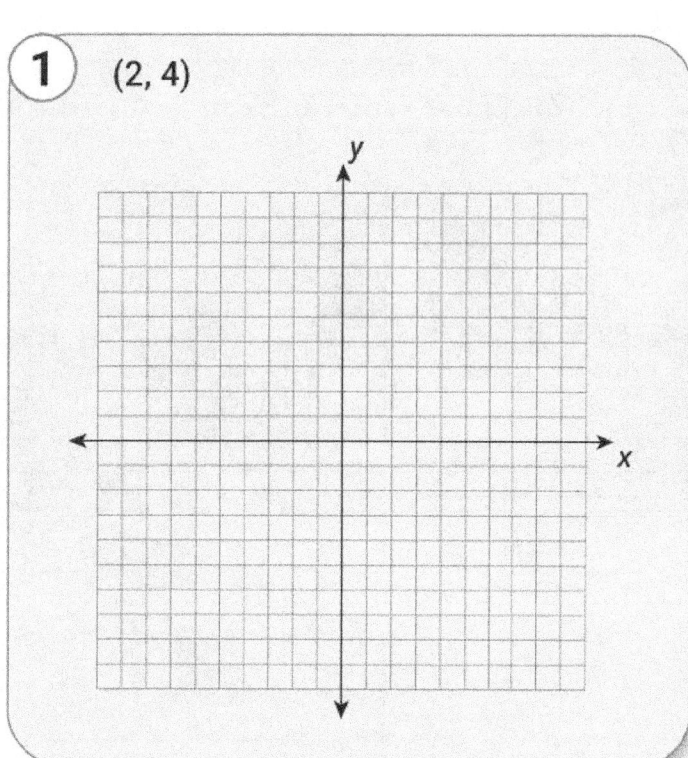

2) (5, 1)

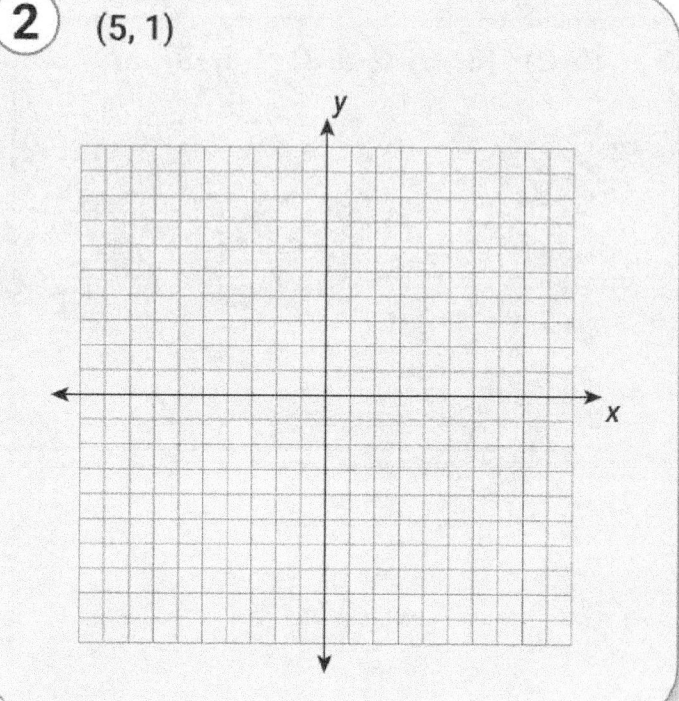

3) (7, 6)

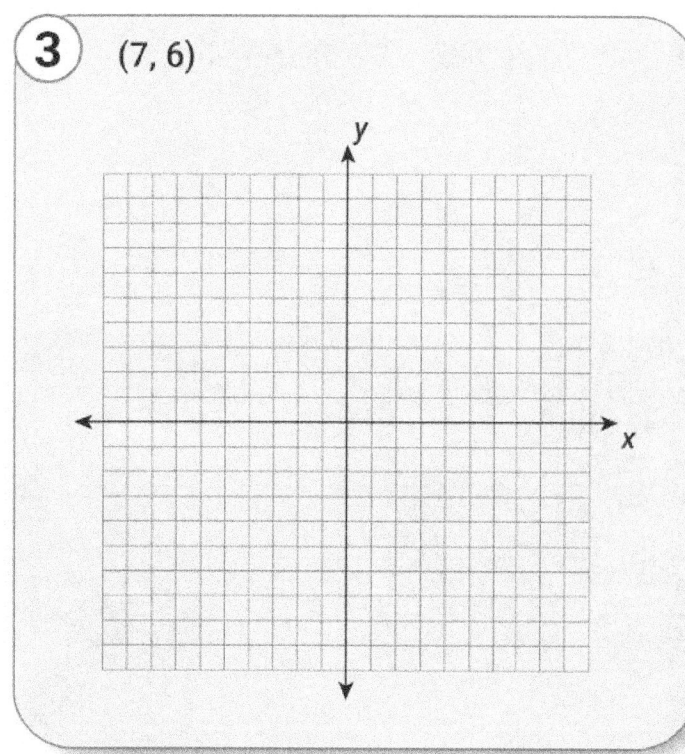

4) (3, -4)

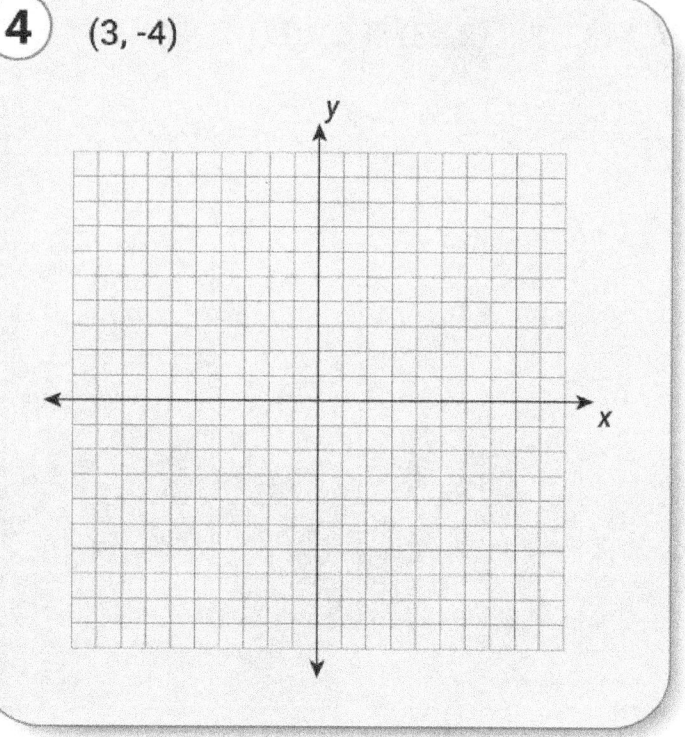

Plot the Point

Plot the point on the provided coordinate plane.

1) (0, 8)

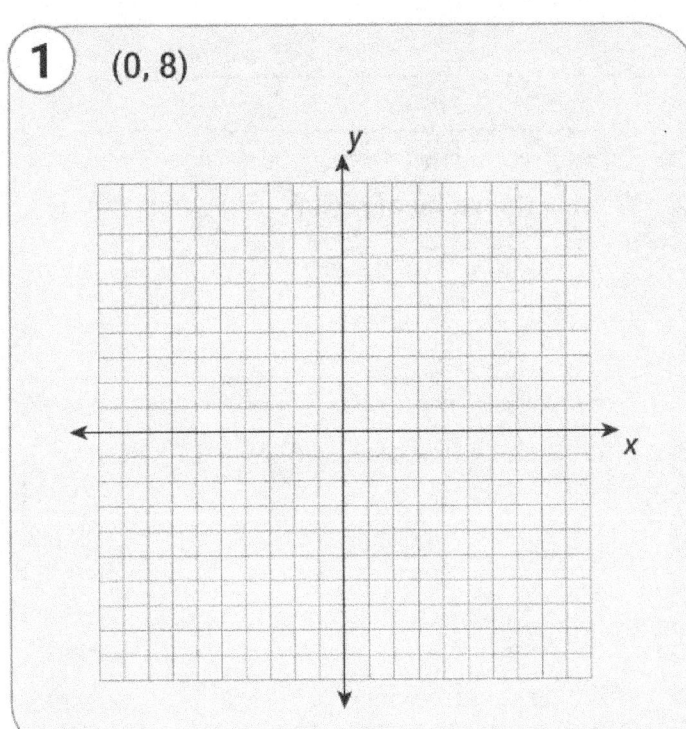

2) (-2, -6)

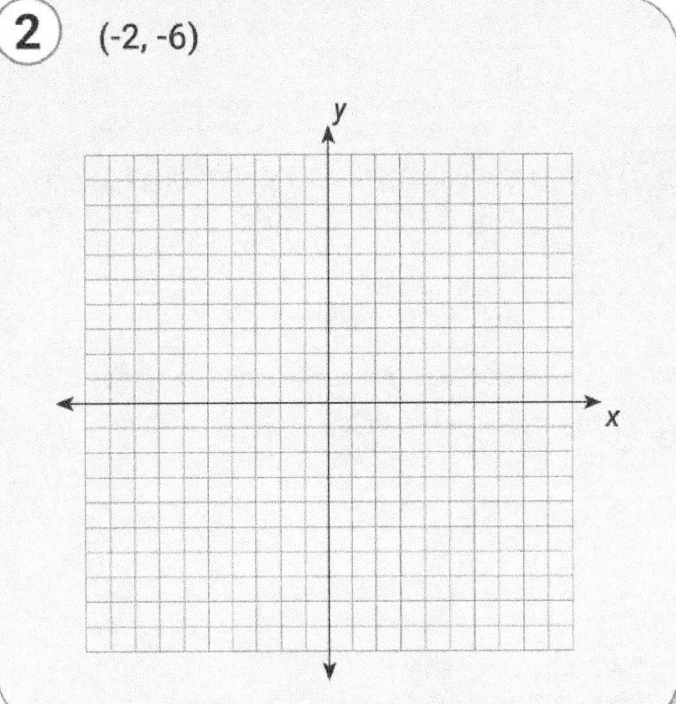

3) (-3, -8)

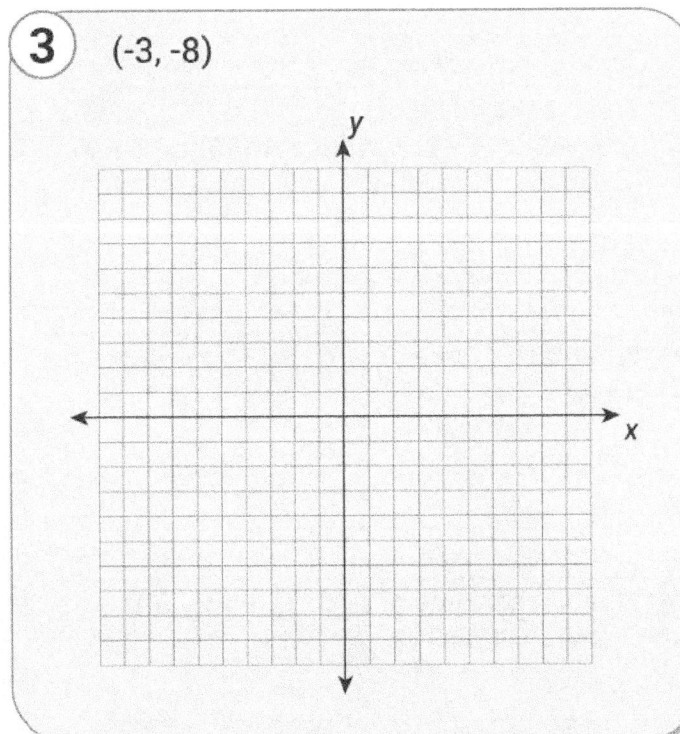

4) (-5, 0)

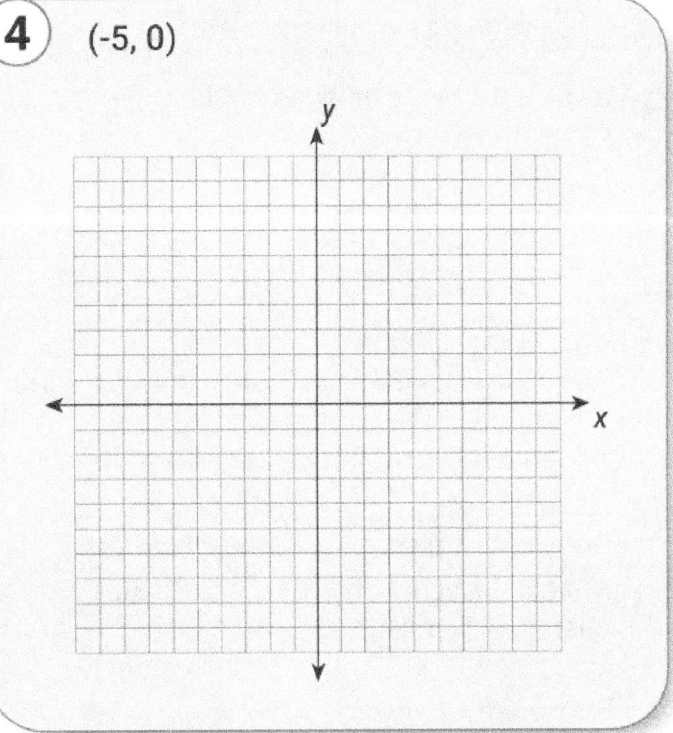

Evaluating Algebraic Expressions

Evaluate the following algebraic expressions for the given values.

1) What is the value of $7b - 2a$ when $a = 8, b = 7$?

2) What is the value of $-3 - 9 - 6c + 7d$ when $c = 4, d = 8$?

3) What is the value of $6x + 7y$ when $x = 5, y = 3$?

4) What is the value of $-2(8m - 6n)$ when $m = 3, n = 6$?

5) What is the value of $x^2 - 2xy + 2y^2$ when $x = 2, y = 3$?

6) What is the value of $8n + 5n^3 + 16n^2$ when $n = 4$?

7) What is the value of $(15 - 8t^2) - (5g^3 - 9 + 6g^2) + (3 + 7t)$ when $t = -2, g = 7$?

8) What is the value of $(2a^2 + 6a^4 - 4a)(3b^3 + 8b^2 + b)$ when $a = 3, b = -6$?

9) What is the value of $9k^2 - 6l^2 + 8k$ when $k = 12, l = -8$?

10) What is the value of $(8b^2 + 3) + (58 - 2b^3) - (6b^3 + 4b)$ when $b = 2$?

11) What is the value of $(8c^3 - 6c^2 + 4) + (3d^3 + 7c^2)$ when $c = -4, d = 3$?

12) What is the value of $y(9 - 7x^4 + 2x)$ when $x = -3, y = 7$?

13) What is the value of $(8 + 4c^2) - (2d^3 - 3d^2)$ when $c = 15, d = 4$?

14) What is the value of $(2m^2 + 3)(4n^2 - 7n)$ when $m = 1, n = -1$?

Solving Equations

Solve each equation for the unknown variable.

1) -5x = -40

2) 2 + j = -8

3) -7 + a = -10

4) -8h + 6h = 22

5) 12 = m − 2

6) 11 = c − 3

7) -12 = 2y

8) 7f = 56

9) -34 = 6.8c

10) 6g = 36

11) $\frac{z}{4} = 7$

12) -5.1d = -35.7

13) $\frac{b}{3} = -4$

14) 2.5t = 10

15) 46.4 = -5.8a

16) $\frac{j}{5} = 4.2$

17) 31.8 = 5.3x

18) -16 = -2b − 4 + 5b

19) 9 = -23u − 22

20) 23 = 13d + 2

Solving Equations

Solve each equation for the unknown variable.

1) $\frac{5}{6}m - 19 = -5$

2) $\frac{1}{7}v + 19 = 21$

3) $7 = \frac{f - 4}{5}$

4) $\frac{2 + 12}{3} = 11$

5) $-7q - 9q = -32$

6) $-10 = \frac{v - 27}{-24}$

7) $\frac{2}{5}n + 21 = 18$

8) $\frac{p - 14}{8} = 28$

9) $-25 = \frac{28 - k}{3}$

10) $4(x - 7) - 8 = 30$

11) $\frac{7 + r}{26} = 23$

12) $2(7 - 3d) = 23$

13) $4x - 3 = 5$

14) $6x + 4 = 16$

15) $\sqrt{(1 + x)} = 4$

16) $\frac{x + 2}{x} = 2$

17) $3x - 4 + 5x = 8 - 10x$

18) $\frac{2x}{5} - 1 = 59$

19) $25 = 5(3 + x)$

20) $12 = 2(3x + 4)$

Adding and Subtracting Polynomials

Simplify each of the following expressions by adding or subtracting the polynomials and collecting like terms.

1) $(9 - 7x^4) - (2x^4 + 3) =$

2) $(8 + 4d^2) - (2d^3 - 6 - 3d^2) =$

3) $(2m^2 + 3) - (4m^2 - 7 + m^4) =$

4) $(x^3 - 3x^2 + 2x - 2) - (3x^3 + 4x - 3) =$

5) $(5x^2 - 3x + 4) - (2x^2 - 7) =$

6) $(7n + 3n^3 + 3) + (8n + 5n^3 + 2n^4) =$

7) $(4 - 8t^2) - (5t^3 - 9 + 6t^2) + (3 + 7t) =$

8) $(2 - 9c^3 + 5c^4) - (8c^4 - 6c^2 + 4) - (3c^3 + 7c^2) =$

9) $(5a^2 - 7a + 2) - (8a - 9) + (4a^4 + 6a^2 + 3) =$

10) $(2a^2 + 6a^4 - 4a) - (3a^2 + 8a^3 + a) + (7a^3 - 5a^4 - 9a^2) =$

11) $(2k - 5k^4) + (7k + 4k^3) - (9k^3 - 6k^4 + 8k) =$

12) $(8b^5 + 3) + (5b^5 + 8 - 2b^3) - (6b^3 + 4b) =$

Multiplying Monomials and Polynomials

Simplify the following expressions by multiplying.

1) $2(7x^2 + 3x - 8) =$

2) $4x^2(-x + 3) =$

3) $6x(7x^2 - 3x + 1) =$

4) $9x^3(2x^2 - 7x - 6) =$

5) $6(3r - 4) =$

6) $12(3x^2 + x) =$

7) $y^2(y + 15) =$

8) $5x(x + 4) =$

9) $(x + 2)(x - 3)(x - 3) =$

Multiplying Monomials and Polynomials

Simplify the following expressions by multiplying.

1. $2x^2(7x^2 - 9x + 2) =$

2. $3x(x^2 + xy + 4y^2) =$

3. $6x^3(3x^2 + 2xy + 5y^2) =$

4. $(4x + 2)^2 =$

5. $(x + 4)^2 =$

6. $(x - 5)(x + 5) =$

7. $(6x - 9)(6x + 9) =$

8. $(8x + 3)^2 =$

9. $(x + 2)(x^2 + 5x - 6) =$

The FOIL Method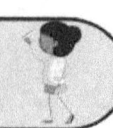

1) $(x - 4)(x + 9) =$ _____

2) $(x - 7)(x - 5) =$ _____

3) $(x + 8)(x + 1) =$ _____

4) $(x + 6)(x + 3) =$ _____

5) $(x - 10)(x - 2) =$ _____

6) $(x + 10)(x + 4) =$ _____

7) $(x - 2)(x + 8) =$ _____

8) $(x - 3)(x + 12) =$ _____

9) $(x + 15)(x - 1) =$ _____

10) $(x + 10)(x + 6) =$ _____

Factoring

Factor each of the following quadratics.

1) $(5a^2 + 5a) = $ _____

2) $t^2 - 3t = $ _____

3) $a^2 + 7a + 12 = $ _____

4) $3w^2 - 12w - 135 = $ _____

5) $(a^2 - 9) = $ _____

6) $(b^2 - 5b) = $ _____

7) $(x^2 - 16) = $ _____

8) $(t^2 + 4t) = $ _____

9) $k^2 - k - 56 = $ _____

10) $16y^2 - 60y + 56 = $ _____

Factoring

Factor each of the following quadratics.

1) $6r^2 + 31r + 40 =$ _____

2) $30v^2 - 105v - 135 =$ _____

3) $6d^2 + 2d - 28 =$ _____

4) $24f^2 + 12f - 36 =$ _____

5) $t^2 + 11t + 18 =$ _____

6) $5p^2 + 5p - 150 =$ _____

7) $40c^2 - 36c - 36 =$ _____

8) $(100y^2 - 36) =$ _____

9) $3g^2 + 12g - 36 =$ _____

10) $(25x^2 - 64) =$ _____

Finding Zeros of Polynomials

Find the zeros of the following polynomials

1) $y = (3x - 2)(x - 3)$
= _____
_____

2) $y = (3x - 5)(x + 5)$
= _____
_____

3) $y = 5x^2 - 39x + 28$
= _____
_____

4) $y = x^3 + 4x^2 + 4x$
= _____
_____

5) $y = (x^2 + x - 20)$
= _____
_____

6) $y = x^3 - 3x^2 - 4x$
= _____
_____

7) $y = (20x^2 - x - 12)$
= _____
_____

8) $y = (10x^2 - 19x - 15)$
= _____
_____

9) $y = (x^3 + 5x^2 - 2x - 24)$
= _____
_____

10) $y = (x^3 + 2x^2 - 9x - 18)$
= _____
_____

11) $y = (x^2 - 7x + 12)$
= _____
_____

12) $y = (25x^2 - 30x + 8)$
= _____
_____

13) $y = (12x^2 - 25x + 12)$
= _____
_____

14) $y = (x^3 + 12x^2 + 47x + 60)$
= _____
_____

15) $y = (25x^2 + 10x - 8)$
= _____
_____

Determining Slope from Coordinate Pairs

Use the two coordinate pairs to calculate the slope of a linear equation that would pass through both points.

1) (-7, -7) and (7, 7)

slope = _____

2) (3, 1) and (-3, -3)

slope = _____

3) (4, 4) and (1, 5)

slope = _____

4) (-5, 2) and (5, 3)

slope = _____

5) (3, -5) and (5, 5)

slope = _____

6) (3, -4) and (5, -2)

slope = _____

7) (2, -5) and (3, 5)

slope = _____

8) (3, 0) and (5, 4)

slope = _____

9) (-1, -4) and (-2, 2)

slope = _____

10) (2, -5) and (3, 5)

slope = _____

Finding Linear Equations from Graphs

Determine the equation of each line graphed on the coordinate plane.

1)

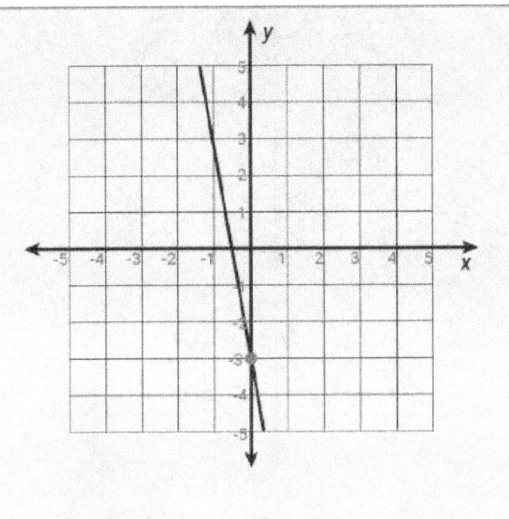

2)

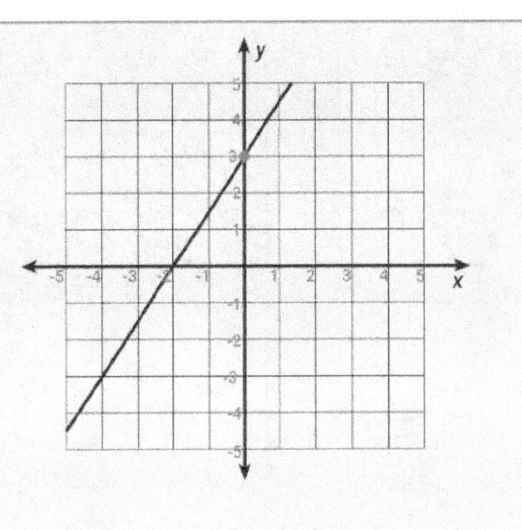

3)

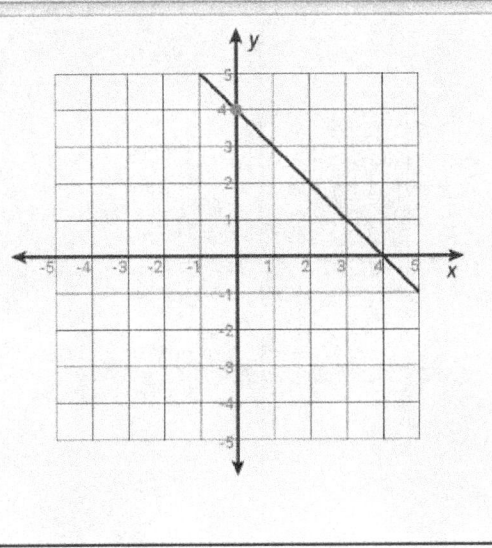

4)

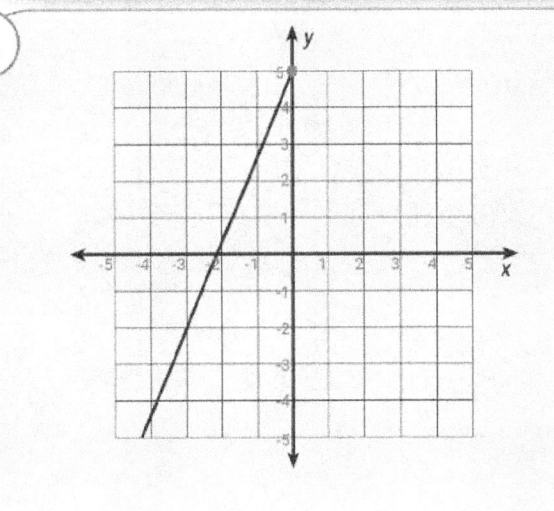

5)

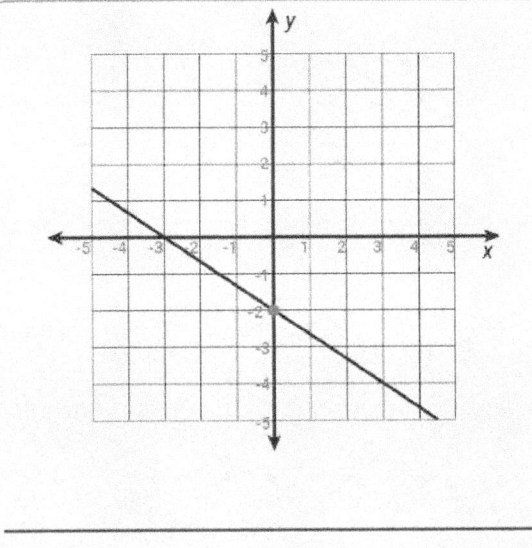

Finding Linear Equations from Graphs

Determine the equation of each line graphed on the coordinate plane.

1)

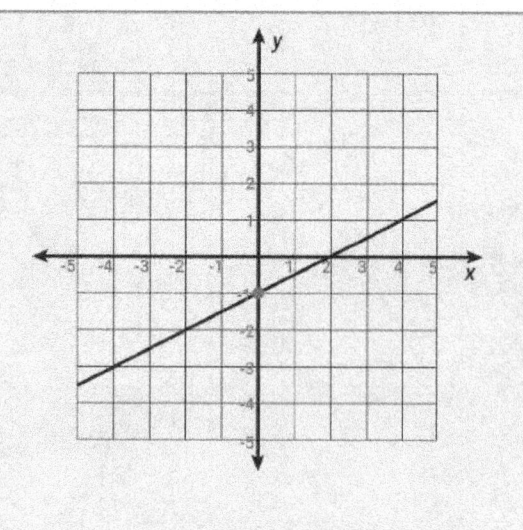

_____

2)

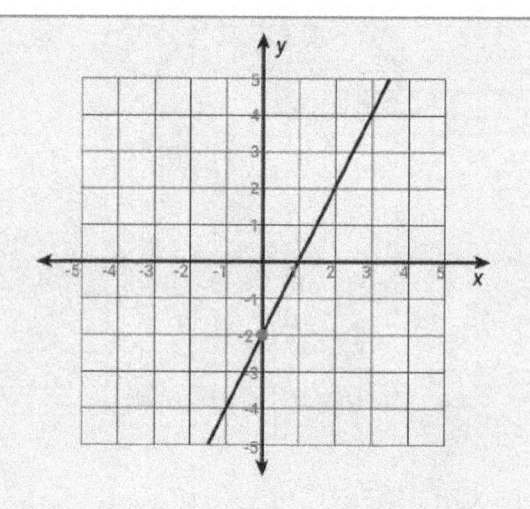

_____

3)

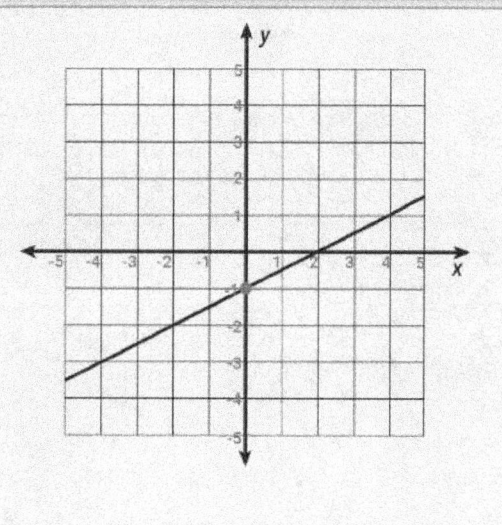

_____

4)

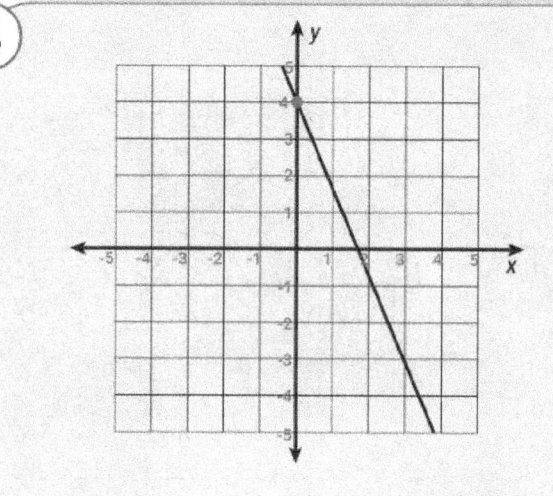

_____

5)

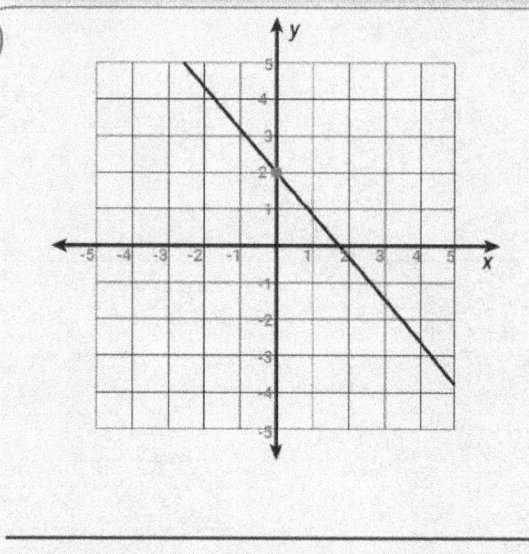

_____

Finding Linear Equations from Graphs

Determine the equation of each line graphed on the coordinate plane.

1.

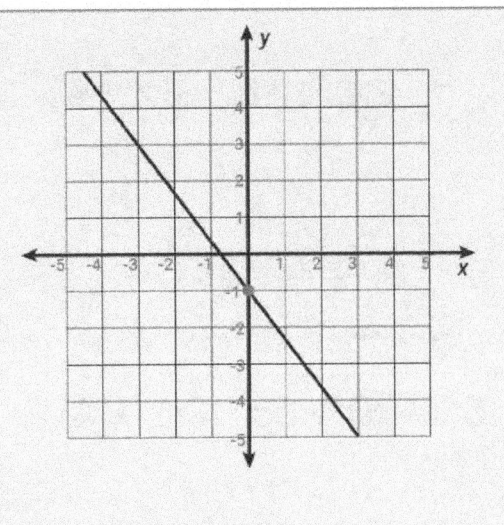

2.

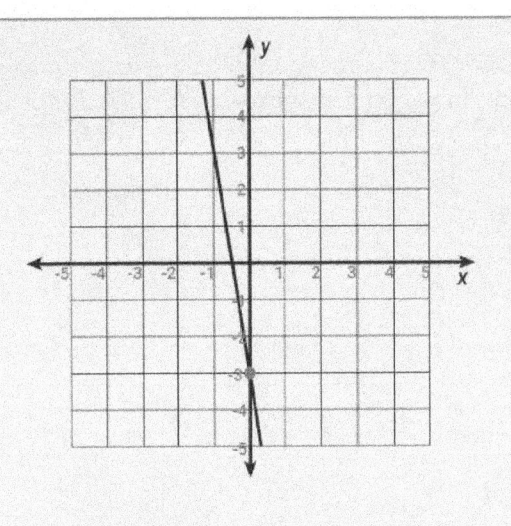

3.

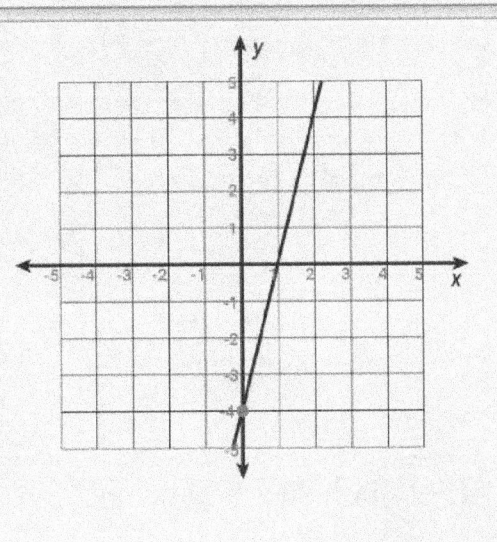

4.

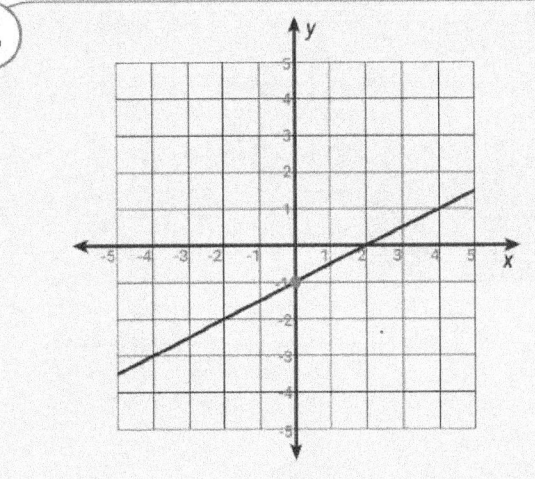

5.

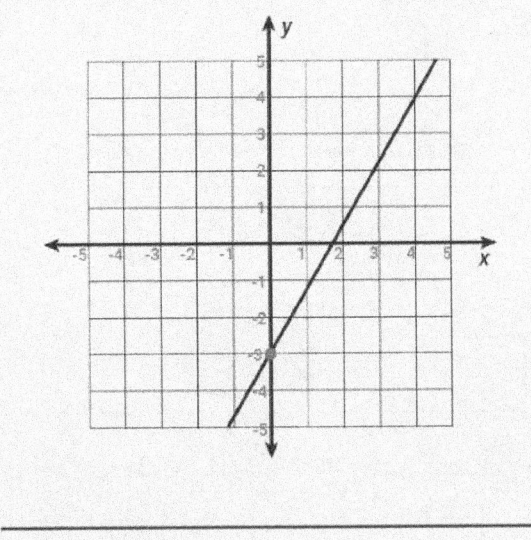

Solving Linear Inequalities

Solve the following inequalities and then graph the solution

1) $4(5 - 3x) < 9x - 64$

2) $2(6 - 2x) \leq 16 - 3(x + 4)$

3) $2(4 - 3x) \geq 8x - 34$

4) $-5(1 - x) > 63 - 3(x + 4)$

5) $8(3 + 2x) - x > -3(x + 4)$

6) $-26 > -3x - 6 + 2x$

7) $48 > 5x - 8 + 2x$

8) $3(6 - 2x) - 4x > 4x - 52$

9) $5x - 146 < 2(3 - 5x) - 4x$

10) $6x - 5 + 3x \geq -77$

Solving Quadratic Equations Exercises

Solve the following quadratic equations using the quadratic formula.

1) $x^2 + 3x - 40 = 0$
= _____

2) $x^2 - 2x - 120 = 0$
= _____

3) $x^2 - 20x - 96 = 0$
= _____

4) $x^2 + 2x - 35 = 0$
= _____

5) $x^2 - 5x - 24 = 0$
= _____

6) $x^2 + 3x - 54 = 0$
= _____

7) $x^2 - 16x + 63 = 0$
= _____

8) $x^2 - 9x + 18 = 0$
= _____

9) $x^2 - 2x - 24 = 0$
= _____

10) $x^2 + 8x + 12 = 0$
= _____

11) $x^2 - 13x + 42 = 0$
= _____

12) $x^2 - 7x - 60 = 0$
= _____

13) $8x^2 - 32x - 40 = 0$
= _____

14) $18x^2 + 72x + 72 = 0$
= _____

15) $12x^2 - 12x - 72 = 0$
= _____

Completing the Square

Solve the following equations by completing the square.

1) $x^2 - 8x - 20 = 0$
= _____

2) $x^2 + 18x + 56 = 0$
= _____

3) $x^2 - 6x - 55 = 0$
= _____

4) $x^2 + 12x - 64 = 0$
= _____

5) $x^2 - 2x - 15 = 0$
= _____

6) $x^2 + 2x = 35$
= _____

7) $x^2 + 2x - 8 = 0$
= _____

8) $x^2 + 10x + 7 = -2$
= _____

9) $x^2 + 8x + 7 = 0$
= _____

10) $x^2 - 19x + 90 = 0$
= _____

11) $4x^2 - 12x - 4 = 12$
= _____

12) $2x^2 - 22x + 60 = 0$
= _____

13) $4x^3 + 8x + 22 = 0$
= _____

14) $100x^2 - 100x - 9 = 0$
= _____

15) $12x^2 - 24x + 6 = -4$
= _____

Solving Systems of Equations with Elimination

Solve the following systems of equations using elimination.

1) $x + 3y = 18$; $-x - 4y = -25$

2) $y = -\frac{3}{2}x - 7$; $y = \frac{1}{2}x + 5$

3) $y = \frac{1}{4}x - 2$; $y = -3x - 15$

4) $3x + y = -21$; $x + y = -5$

5) $-5x - 7y = 24$; $10x + 7y = 1$

6) $5x - 2y = 18$; $-2x - y = -9$

7) $y = -\frac{2}{7}x - 2$; $y = -\frac{1}{7}x - 4$

8) $3x + 3y = -3$; $-3x - 4y = 2$

9) $-x + 2 = y$; $-6x - 3 = y$

10) $4x - y = -1$; $-3x + 2y = -3$

Solving Systems of Equations with Substitution

1) $2x - 3y = -1; -x + y = -1$

2) $y = \frac{1}{2}x + 4; y = -\frac{5}{2}x + 10$

3) $-16 = 8x + y; -3x + y = -5$

4) $y = 4x - 10; y = \frac{1}{3}x + 1$

5) $y = 4x + 5; y = -\frac{1}{3}x - 8$

6) $y = -4x + 15; y = -\frac{7}{2}x + 12$

7) $5x + 2y = 21; -x - y = -9$

8) $6x - 5y = 12; 2x + y = 20$

9) $5 = 4x - 7y; 9x - 7y = -15$

10) $y = -\frac{7}{5}x - 3; y = -\frac{4}{9}x - 3$

Functions

Use the provided functions to complete the missing values in the input/output tables below.

1) $f(x) = x^2 - 3$

x	f(x)
-2	
	-2
0	
1	
2	

2) $f(x) = 2x + 4$

x	f(x)
-2	
	2
1	
	8
	14

3) $f(x) = 5x^2 - 1$

x	f(x)
	44
-1	
2	
	44
5	

4) $f(x) = 8 - 5x$

x	f(x)
-6	
-3	
0	
	-7
	-17

5) $f(x) = -3x + 6$

x	f(x)
	21
-1	
	6
	-9
	-15

Functions

Use the provided functions to complete the missing values in the input/output tables below.

1) $f(x) = x^2 - 2$

x	f(x)
-4	
-3	
0	
1	
2	

2) $f(x) = 3 - 2x$

x	f(x)
	9
	5
	3
	-1
	-5

3) $f(x) = 2x^3$

x	f(x)
-2	
-1	
1	
2	
3	

4) $f(x) = \dfrac{x}{3}$

x	f(x)
	-2
	-1
	1
	2
	3

5) $f(x) = \dfrac{x^3}{2}$

x	f(x)
-2	
	0.5
	4
	13.5
	32

Algebraic Word Problems

Solve each of the following word problems.

1) Store brand coffee beans cost $1.23 per pound. A local coffee bean roaster charges $1.98 per $1\frac{1}{2}$ pounds. How much more would 5 pounds from the local roaster cost than 5 pounds of the store brand?

2) Paint Inc. charges $2000 for painting the first 1,800 feet of trim on a house and $1.00 per foot for each foot after. How much would it cost to paint a house with 3125 feet of trim?

3) Sam is twice as old as his sister, Lisa. Their oldest brother, Ray, will be 25 in three years. If Lisa is 13 years younger than Ray, how old is Sam?

4) A rectangle has a length that is 5 feet longer than three times its width. If the perimeter is 90 feet, what is the length in feet?

5) Carly purchased 84 bulbs for her flower garden. Tulips came in trays containing six bulbs and daffodils came in trays containing 8 bulbs. Carly bought an equal number of tulip and daffodil trays. How many of each type of flower bulb were purchased?

6) Apples cost $2 each, while bananas cost $3 each. Maria purchased 10 fruits in total and spent $22. How many apples did she buy?

7) Jessica buys 10 cans of paint. Red paint costs $1 per can and blue paint costs $2 per can. In total, she spends $16. How many red cans did she buy?

Algebraic Word Problems

Solve each of the following word problems.

1) Cameron sold half his yo-yo collection then bought six more. He now has 16 yo-yos. How many did he begin with?

2) The sum of three consecutive pages in a book is 144. What is the lowest page number in the set?

3) Jane got 9 tickets from winning a game at the arcade and adds them to her pile. She then spends half of her tickets on a mini football, leaving just 26 left. How many tickets did she have before buying the football?

4) Noah had $165 to spend on 8 identical posters to give as favors at his birthday party. After buying the posters, he had 29 dollars left. How much was each poster?

5) Three-hundred-thirty-one third grade students are taking a field trip to the local science museum. They are filling 7 buses and then 9 students are riding in an accessible van. How many students fit on each bus?

6) After seeing a movie, Dennis, Jan, and Tina decide to split the total cost of the tickets and the snacks. If they each had to pay $13, and the snacks cost $12 total, how much was each ticket?

7) Tiara is selling cups of lemonade for $0.75. If each pitcher makes 8 cups of lemonade and the cost of ingredients is $1.12 per pitcher, how many cups does she need to sell to make $20 profit?

Algebraic Word Problems

Solve each of the following word problems.

1) Peter and his sister, Lisa, are collecting bottles and cans for redemption and plan to donate the proceeds to the local animal shelter. The redemption center gives $0.05 for each aluminum can, but glass bottles earn $0.10. If they bring their bottles and cans to the redemption center and get $29.00 total after redeeming ¾ as many glass bottles as cans, how many bottles did they redeem?

2) Mei is baking brownies and cookies for a bake sale. The recipe for brownies makes 16 brownies and the recipe for cookies makes 24 cookies. How many batches of cookies does she need to make if she is to bring 184 treats and she makes one more batch of cookies than brownies?

3) Dwayne's baseball team has 44 games on the schedule this season. If ¼ are home games, how many games are away?

4) Juanita is counting ladybugs and spiders that she finds in her backyard. Altogether, she counts 198 legs. If she sees twice as many spiders as ladybugs, how many ladybugs were there?

5) Mr. Read's seventh grade class is studying geography. If he is creating equal-sized groups to present on each of the seven continents and each group has four students, how many students are in the class?

6) Shankar is training for a 10k road race. If he runs 5 miles in 42:30, then how long will it take to run 7 miles if he maintains the same pace?

7) Becca practices piano 7 days a week. Some days, she plays 30 minutes, and some days, she plays 50 minutes. If she played 5 hours and 10 minutes in the week, how many days did she play 50 minutes?

Algebraic Word Problems

Solve each of the following word problems.

1) Sam's mom works at the local diner. She makes $8 per hour as a base wage plus tips. Last week, she earned $396. If tips made up 1/3 of her pay, how many hours did she work?

2) Valencia babysits a family with 3 kids. She makes $12 per hour. However, when one of the kids had a friend over, she makes an extra $3 per hour. Last month, she babysat 6 times. If one of those times there was an additional child, how much did she make if each occasion was 4 hours (including the one with a friend), except one occasion, which was 6 hours?

3) The soccer team is selling donuts to raise money to buy new uniforms. For every box of donuts that they sell, the team receives $3 towards their new uniforms. There are 15 people on the team. How many boxes does each player need to sell in order to raise $270 for their new uniforms?

4) At the store, Jan spends $90 on apples and oranges. Apples cost $1 each and oranges cost $2 each. If Jan buys the same number of apples as oranges, how many oranges did she buy?

5) Kristen purchases $100 worth of CDs and DVDs. The CDs cost $10 each and the DVDs cost $15. If she bought four DVDs, how many CDs did she buy?

6) In Jim's school, there are 3 girls for every 2 boys. There are 650 students in total. How many students are girls?

7) Kimberley earns $10 an hour babysitting, and after 10 p.m., she earns $12 an hour, with the amount paid being rounded to the nearest hour accordingly. On her last job, she worked from 5:30 p.m. to 11 p.m. In total, how much did Kimberley earn for that job?

Points, Lines, and Planes

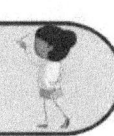

Use the image below to answer questions 1 – 6:

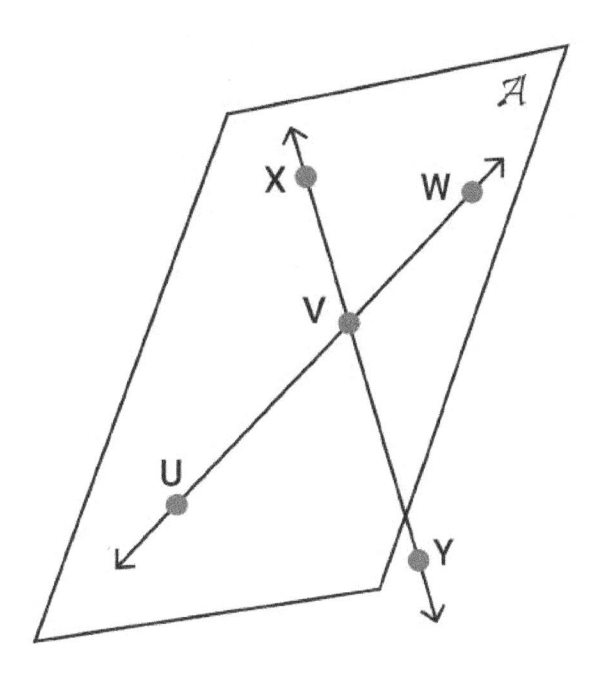

1. Name any 3 points _____ _____ _____

2. Name 2 line segments _____ _____

3. Name a set of 3 points that are collinear and coplanar _____ _____ _____

4. Name 2 lines _____ _____

5. Name a point that is not coplanar with X and V _____

6. Name 2 rays _____ _____

Points, Lines, and Planes

Use the image below to answer questions 1 – 6:

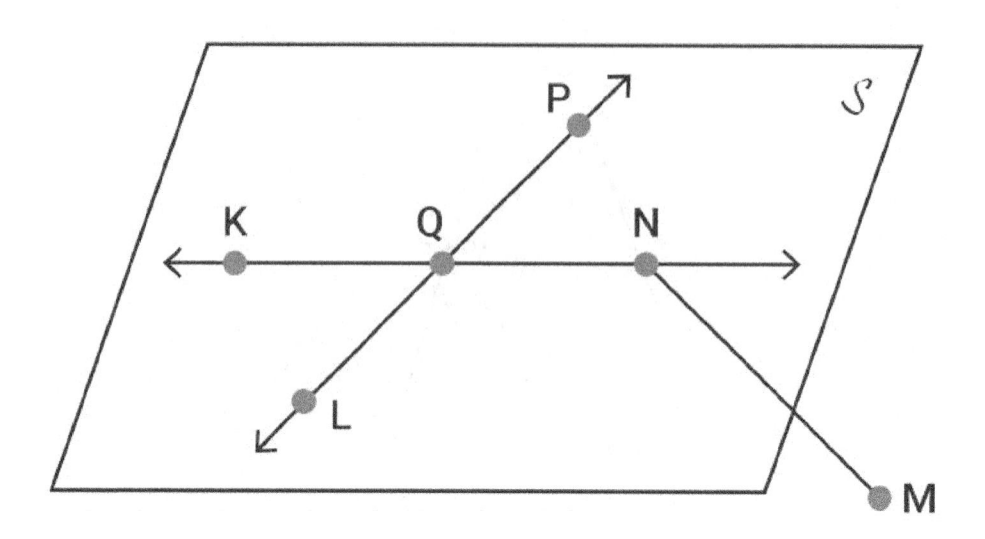

1 Name a point that is not collinear with L but is coplanar _____

2 Name a line segment that could not also be named as a ray or line _____

3 Name 2 points that are collinear with L _____ _____

4 Name 2 lines _____ _____

5 Name a point that is not coplanar with P and Q _____

6 Name 2 rays _____ _____

Points, Lines, and Planes

Use the image below to answer questions 1 – 7:

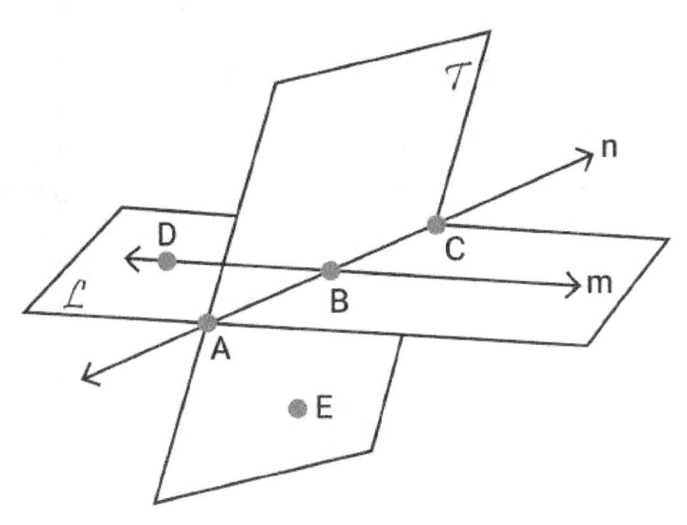

1 Name 2 planes _____ _____

2 Name a point that is collinear with D _____

3 Name a point that is not collinear with C _____

4 Write another way to name line n _____

5 Name a pair of points that are not collinear in plane T _____ _____

6 Name the line where L and T intersect _____

7 Name a point that is coplanar with D _____

Points, Lines, and Planes

Use the image below to answer questions 1 – 10:

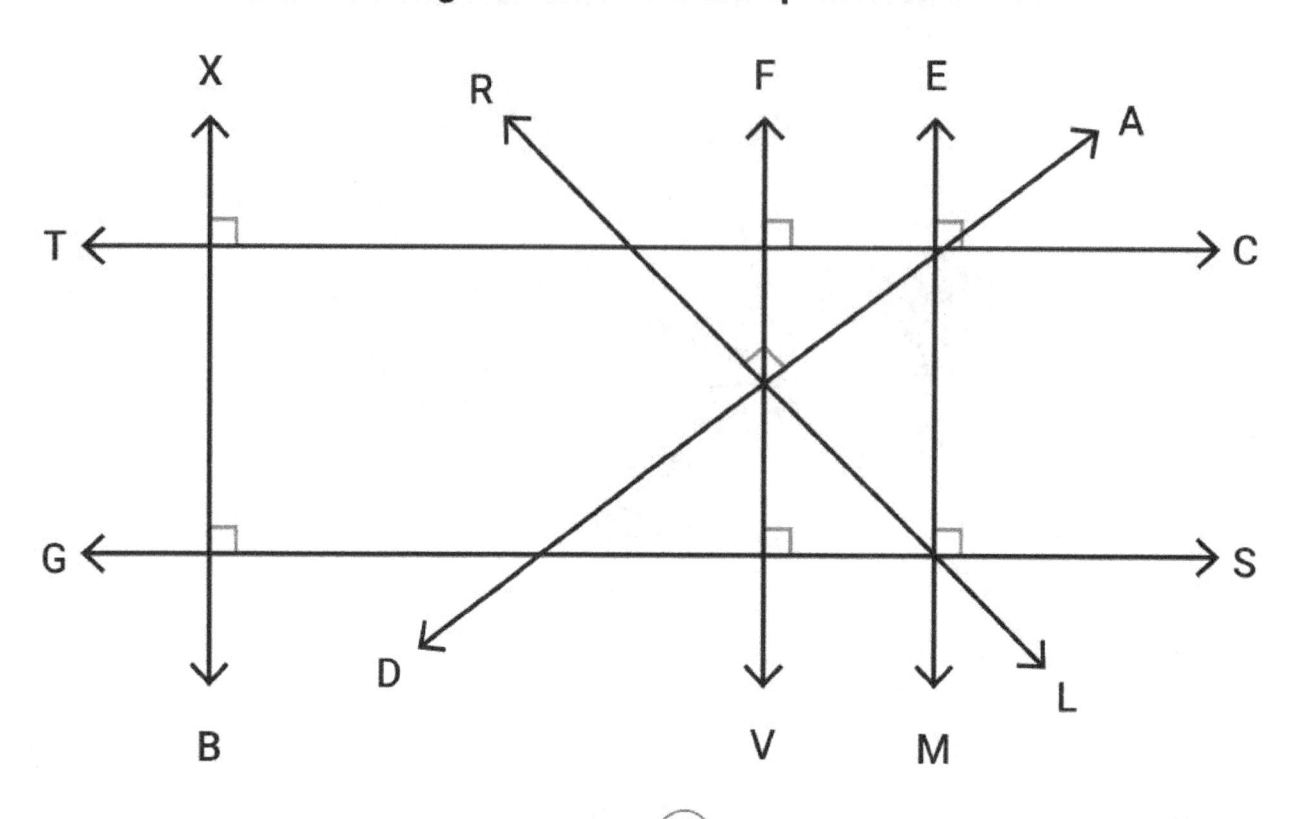

1. Lines EM and FV are _____ lines.

2. Lines GS and TC are _____ lines.

3. Lines AD and LR are _____ lines.

4. Lines BX and EM are _____ lines.

5. Lines AD and GS are _____ lines.

6. Lines BX and FV are _____ lines.

7. Lines EM and AD are _____ lines.

8. Lines BX and GF are _____ lines.

9. Lines TC and FV are _____ lines.

10. Lines RL and FV are _____ lines.

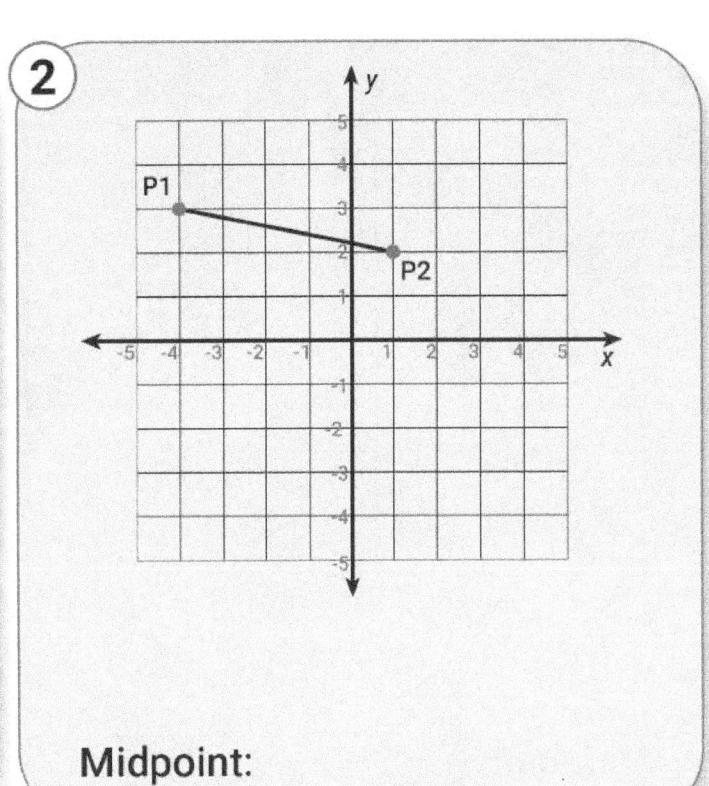

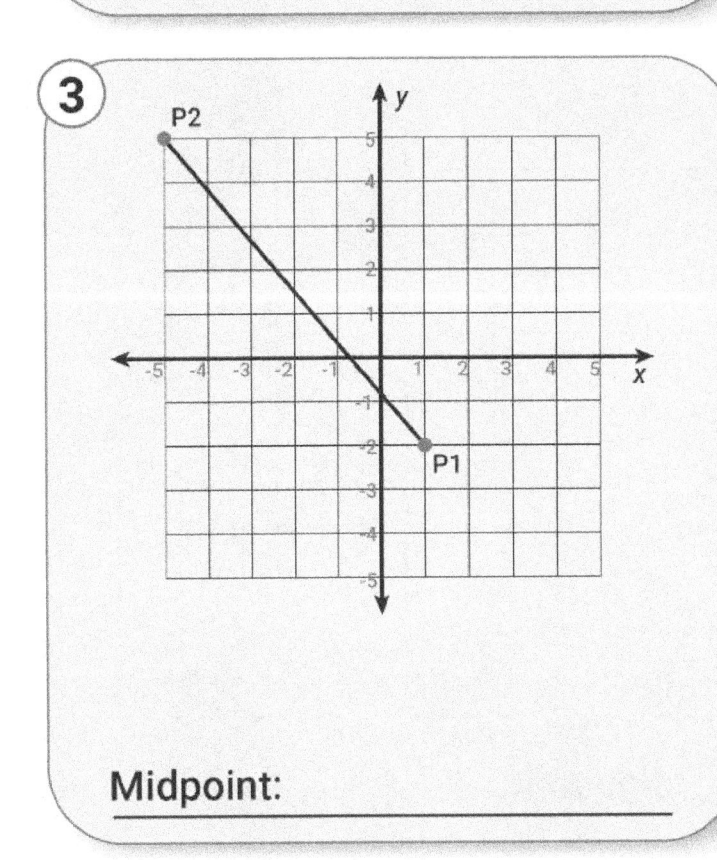

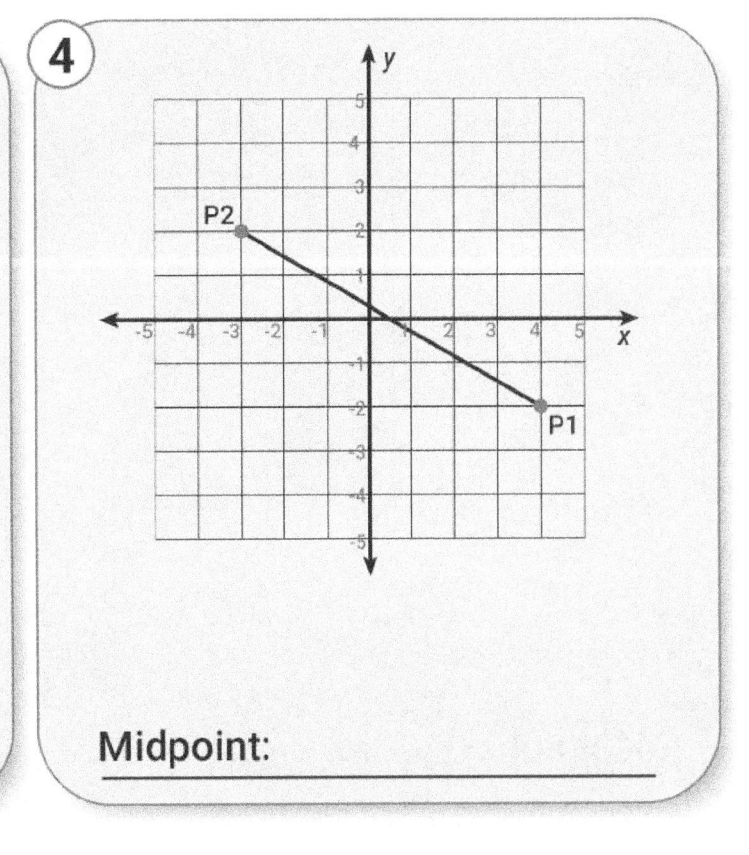

Where is the Middle?

Determine the midpoint of the given line segments and plot it on the line.

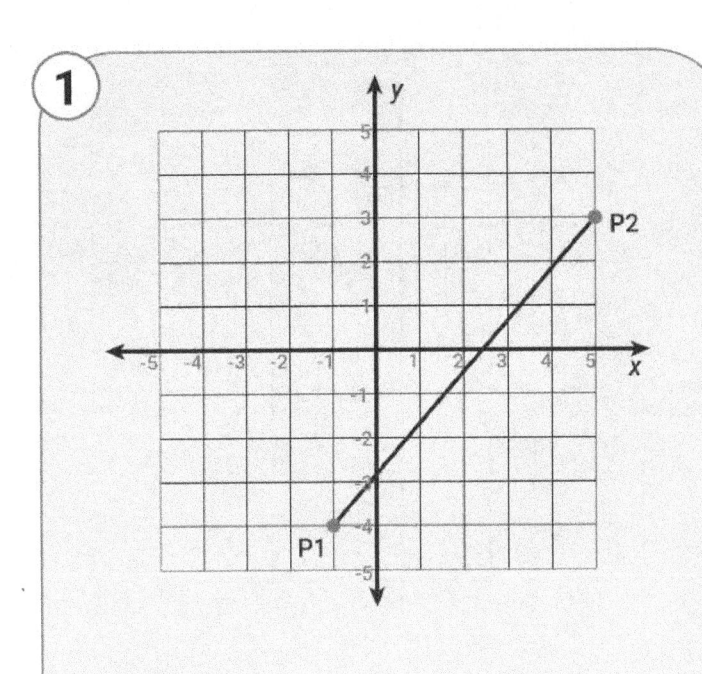

1

P1 = (-1, -4), P2 = (5, 3)

Midpoint: _____

2

P1 = (-4, 3), P2 = (2, 2)

Midpoint: _____

3

P2 = (-5, 5), P1 = (1, -2)

Midpoint: _____

4

P2 = (-3, 2), P1 = (4, -2)

Midpoint: _____

Where is the Middle?

Determine the midpoint of the given line segments and plot it on the line.

1)

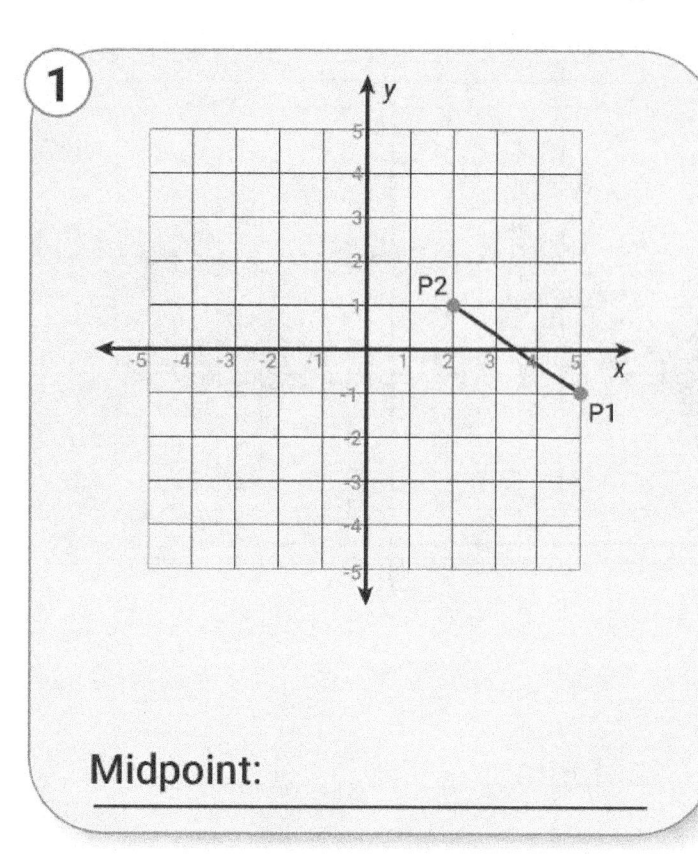

P2 (2, 1), P1 (4, -1)

Midpoint: _____

2)

P1 (1, 3), P2 (5, 2)

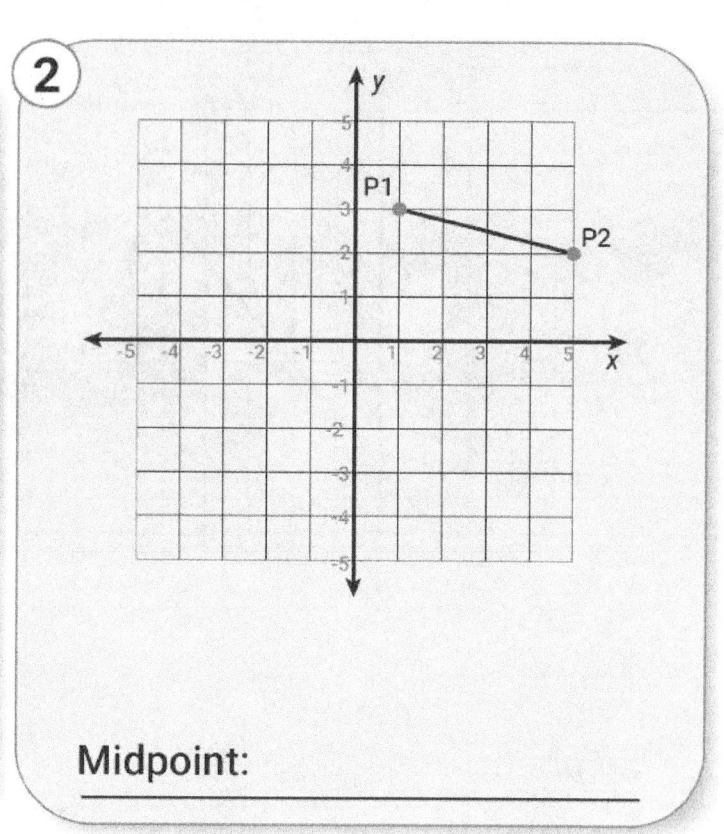

Midpoint: _____

3)

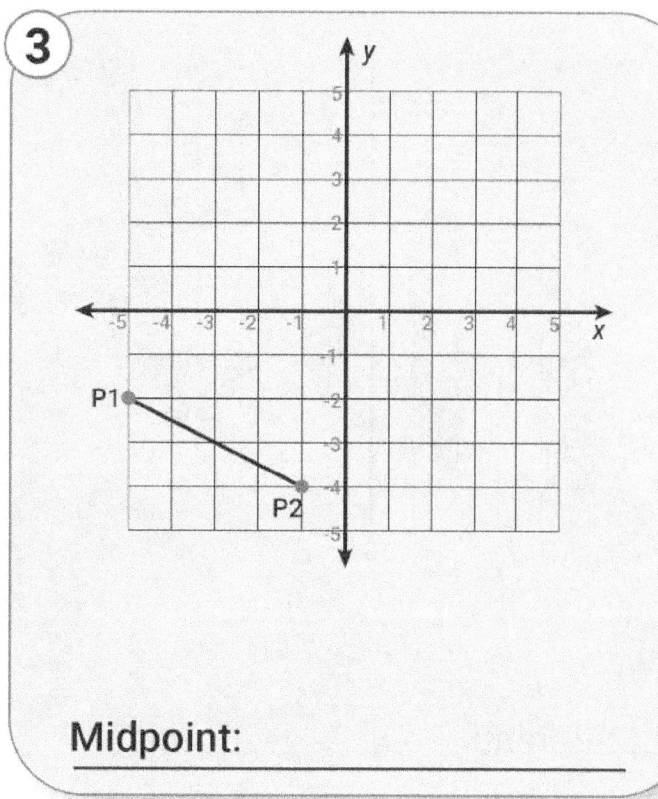

P1 (-5, -2), P2 (-2, -4)

Midpoint: _____

4)

P2 (-4, 3), P1 (5, 2)

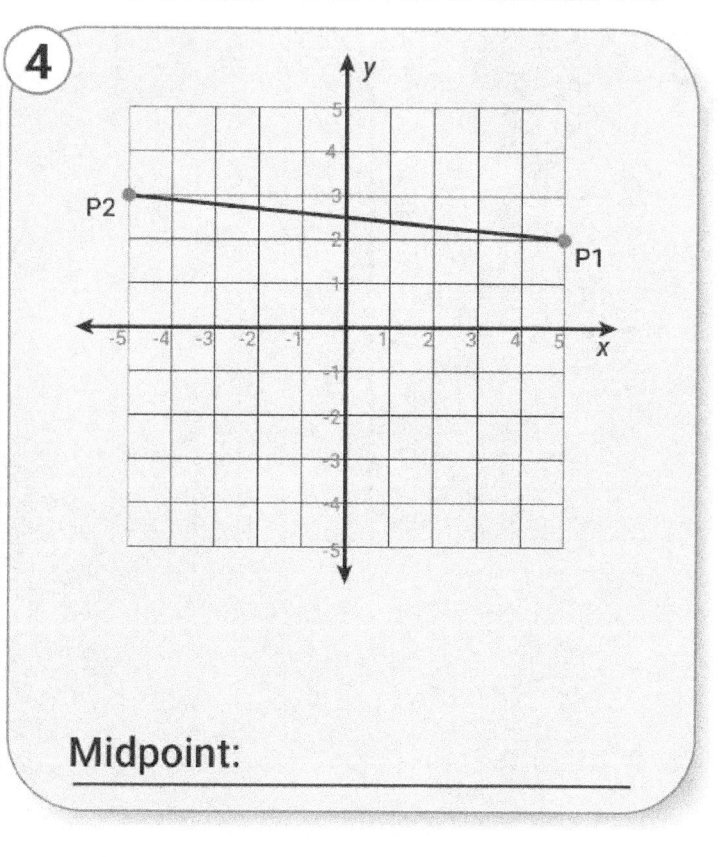

Midpoint: _____

Where is the Middle?

Determine the midpoint of the given line segments and plot it on the line.

1

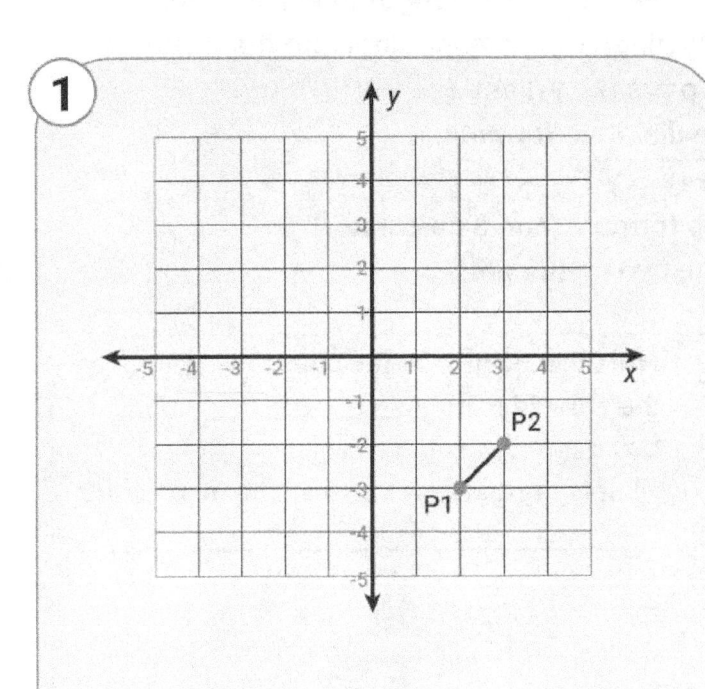

P1 (2, -3), P2 (3, -2)

Midpoint: _____

2

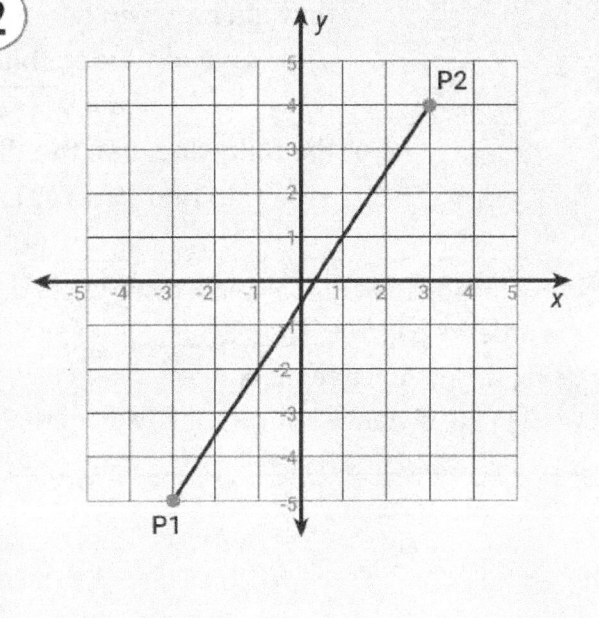

P1 (-1, -5), P2 (4, 4)

Midpoint: _____

3

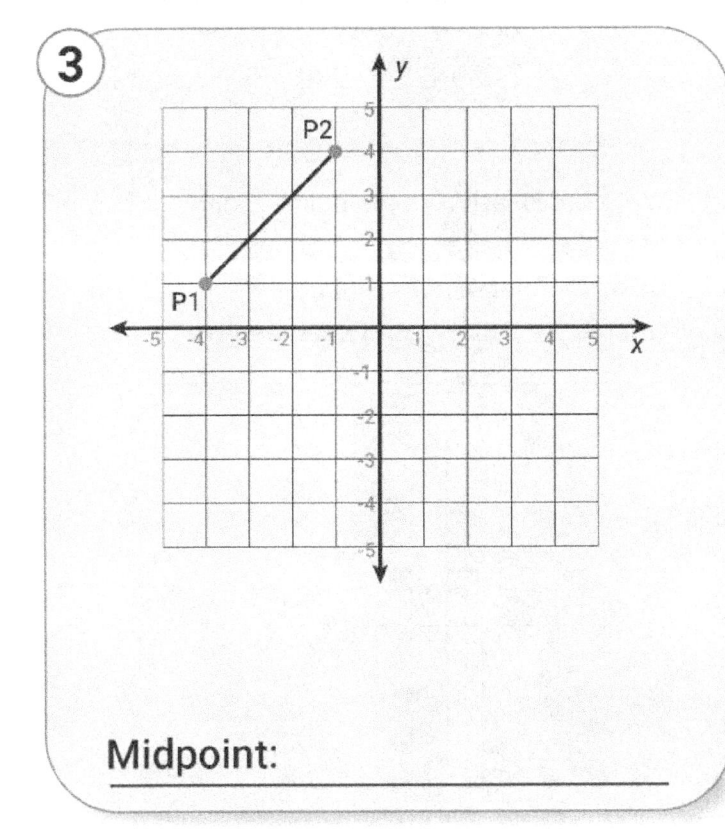

P1 (-4, 1), P2 (-1, 4)

Midpoint: _____

4

P1 (4, 3), P2 (-1, -3)

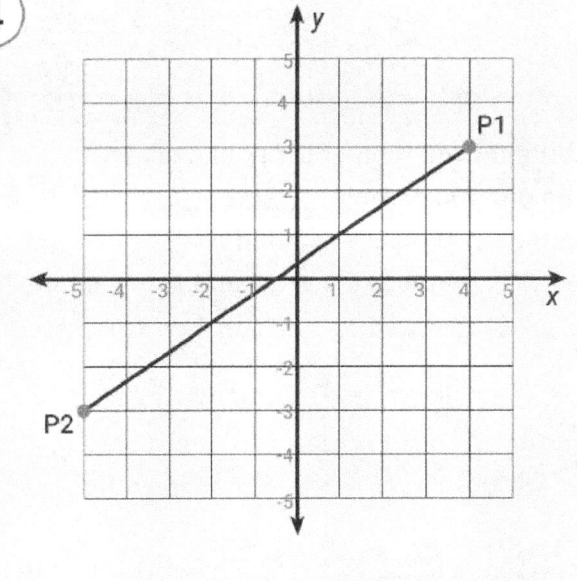

Midpoint: _____

How Far Apart?

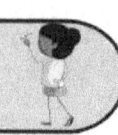

The Pythagorean theorem states that a² + b² = c².
The side lengths of a shape plotted on the coordinate plane, or the length of a line segment, or the distance between two points can be found by plugging the endpoints into the distance formula between two ordered pairs (x_1, y_1) and (x_2, y_2).
As a reminder, this is the **distance formula**:
$$d = \sqrt{(x_2 - x_1)^2 + (y_2 - y_1)^2}$$
For the following, use the distance formula (and a calculator!)
to find how far apart the two points are.

1) Write the location of the points using the ordered pairs: A: _____ ; B: _____
Using the distance formula, what is the distance between the two points?
_____

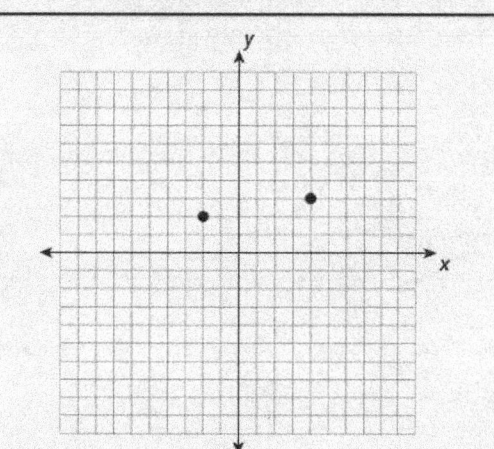

2) Write the location of the points using the ordered pairs: A: _____ ; B: _____
Using the distance formula, what is the distance between the two points?
_____

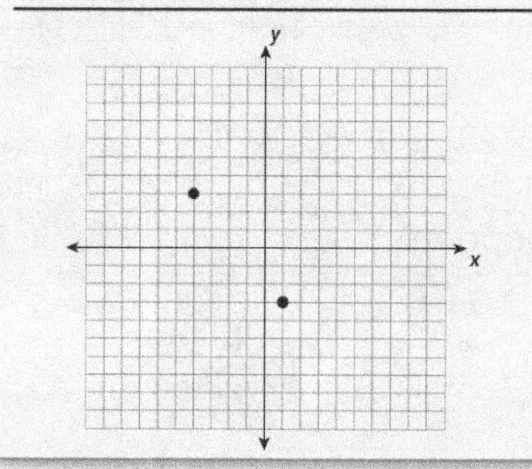

3) Write the location of the points using the ordered pairs: A: _____ ; B: _____
Using the distance formula, what is the distance between the two points?
_____

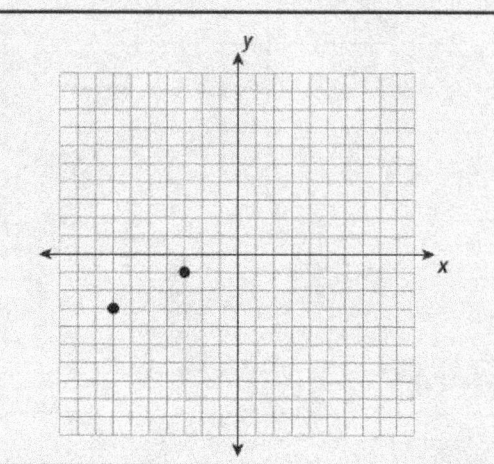

4) Write the location of the points using the ordered pairs: A: _____ ; B: _____
Using the distance formula, what is the distance between the two points?
_____

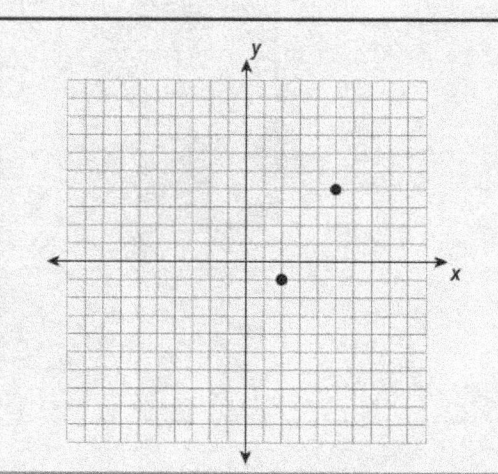

Determining the Measurement of Missing Angles

Find the missing measures of the indicated angles.

1) What is the measure of angle ABD if angle ABC = 55°?

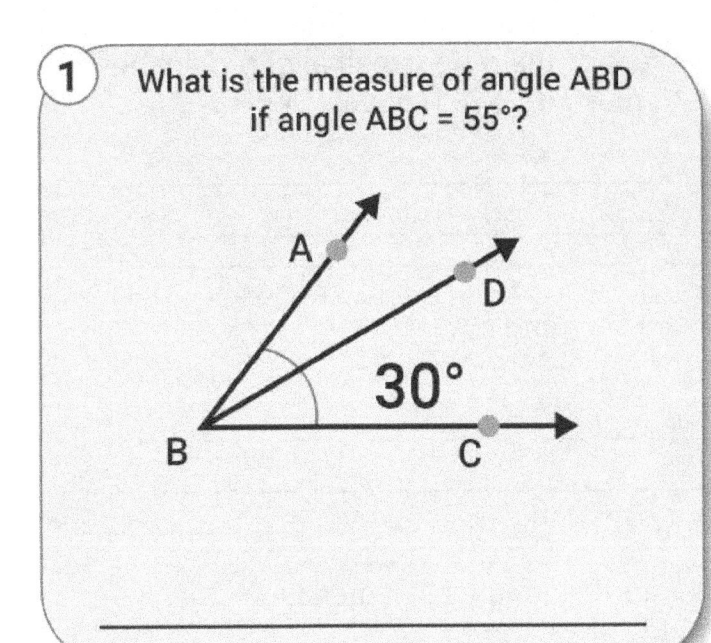

30°

_____

2) What is the measure of angle GHJ if angle GHK = 120°?

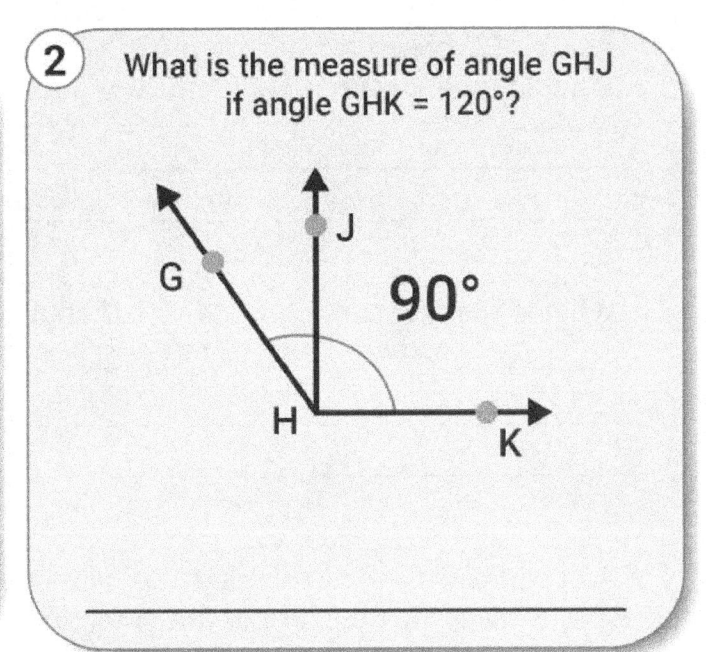

90°

_____

3) What is the measure of angle ABC if angle ABD is 97°?

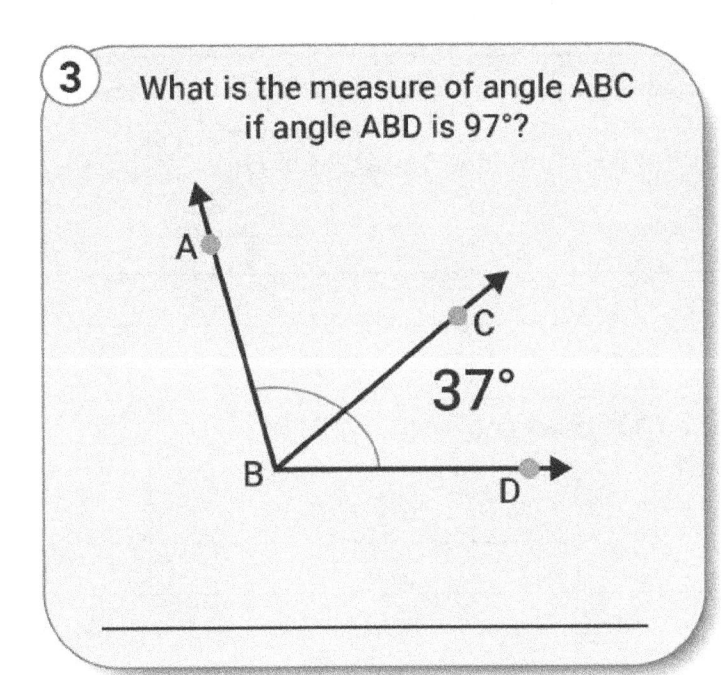

37°

_____

4) What is the measure of angle RSU if angle TSU measures 65°?

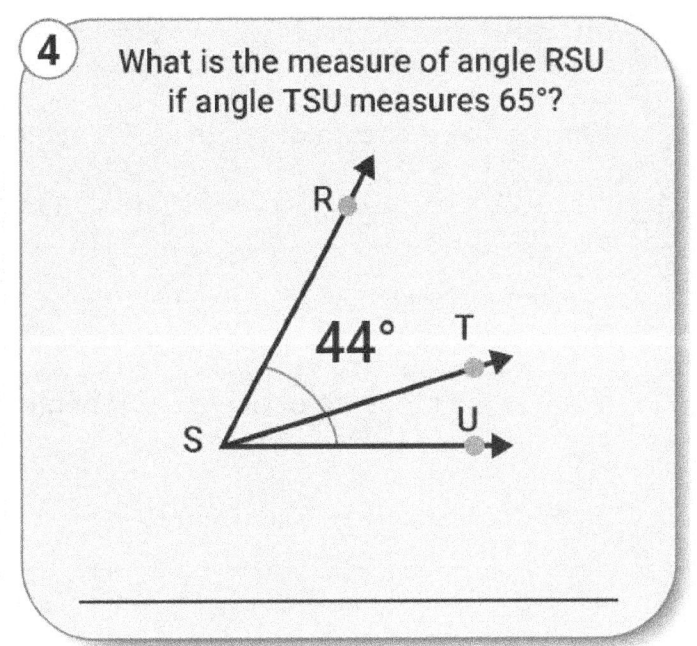

44°

_____

Determining the Measurement of Missing Angles

Find the missing measures of the indicated angles.

5. Angle ACD measures 116°. It is divided into two smaller angles:
Angle ABC measures 48°. Angle BCD, the measure of the other small angle, is unknown.
What is the measure of angle BCD if angle ABC plus angle BCD equals angle ACD?

6. What is the measure of angle VYZ if angle XYZ = 23°, and angle XYV = 11°?

7. Angle LMO = 84°. It is divided into two smaller angles: Angle LMN measures 35°.
Angle MNO, the measure of the other small angle, is unknown.
What is the measure of angle MNO if angle LMN plus MNO equals angle?

8. What is the measure of angle CEF if angle DEF = 91°, and angle CED = 48°?

9. What is the measure of angle IJK if angle HIJ = 12°, and angle HIK = 27°?

10. What is the measure of angle GHJ if angle GHK = 125° and angle JHK is 92°?

Using Angle Relationships to Determine the Missing Measures

Use your knowledge of angle relationships
to determine the measurements of the angles in the figure.
Note that the figures may not be drawn to scale.

1) What are the measures of angles x and y in the image below?

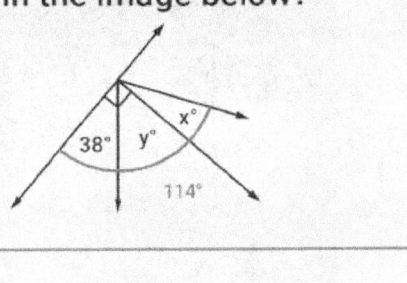

(Figure: angles 38°, y°, x°, 114°)

_____
_____
_____

2) What is the measure of angle x in the image below?

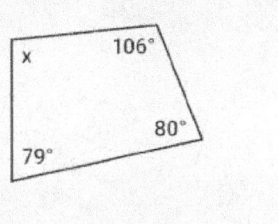

(Figure: quadrilateral with angles x, 106°, 80°, 79°)

_____
_____
_____

3) ∠2: _____, ∠3: _____, ∠4: _____,
∠5: _____, ∠6: _____, ∠7: _____,
∠8: _____,

(Figure: two parallel lines cut by a transversal; ∠1 = 75°, with ∠2, ∠3, ∠4, ∠5, ∠6, ∠7, ∠8)

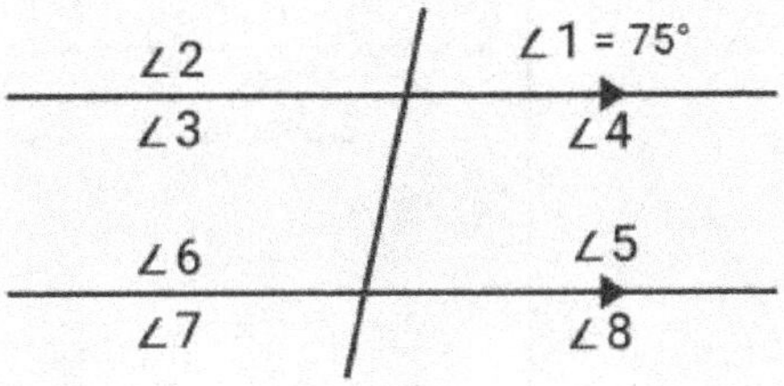

4) What is the measure of angle x in the image below?

(Figure: two triangles meeting at a vertex; 140°, y°, x°, z°, 40°)

_____
_____

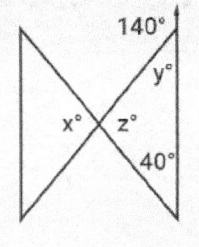

5) What is the measure of angle x in the image below?

(Figure: right triangle with x°, y°, z°, 115°)

_____
_____
_____
_____
_____

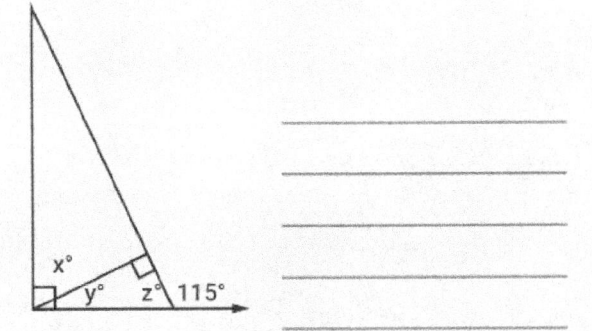

Using Angle Relationships to Determine the Missing Measures

Use your knowledge of angle relationships to determine the measurements of the angles in the figure. Note that the figures may not be drawn to scale.

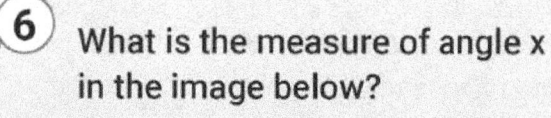

6) What is the measure of angle x in the image below?

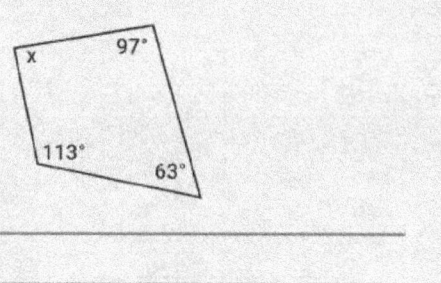

7) What is the measure of angle y in the image below?

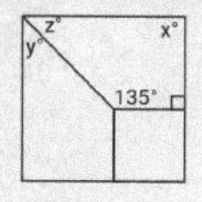

8) What are the measurements of angles x, y, and z in the image below? Note that the internal shape is a regular pentagon.

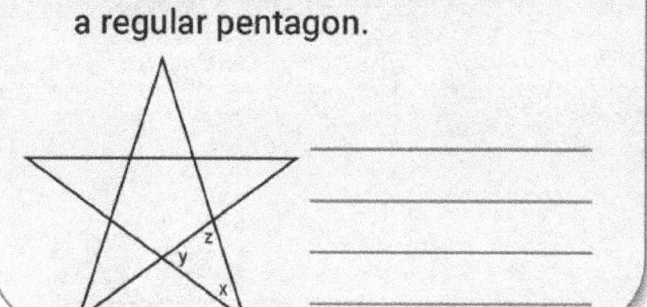

9) What is the measure of angle y in the image below?

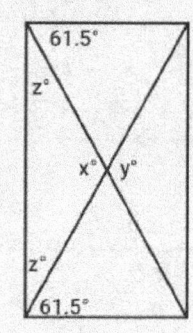

10) What is the measure of angle x in the image below?

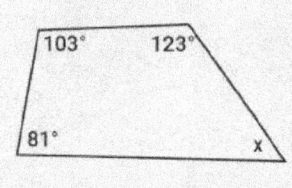

Classifying Triangles

Classify the following triangles by their sides and angles
(for example, isosceles and right).

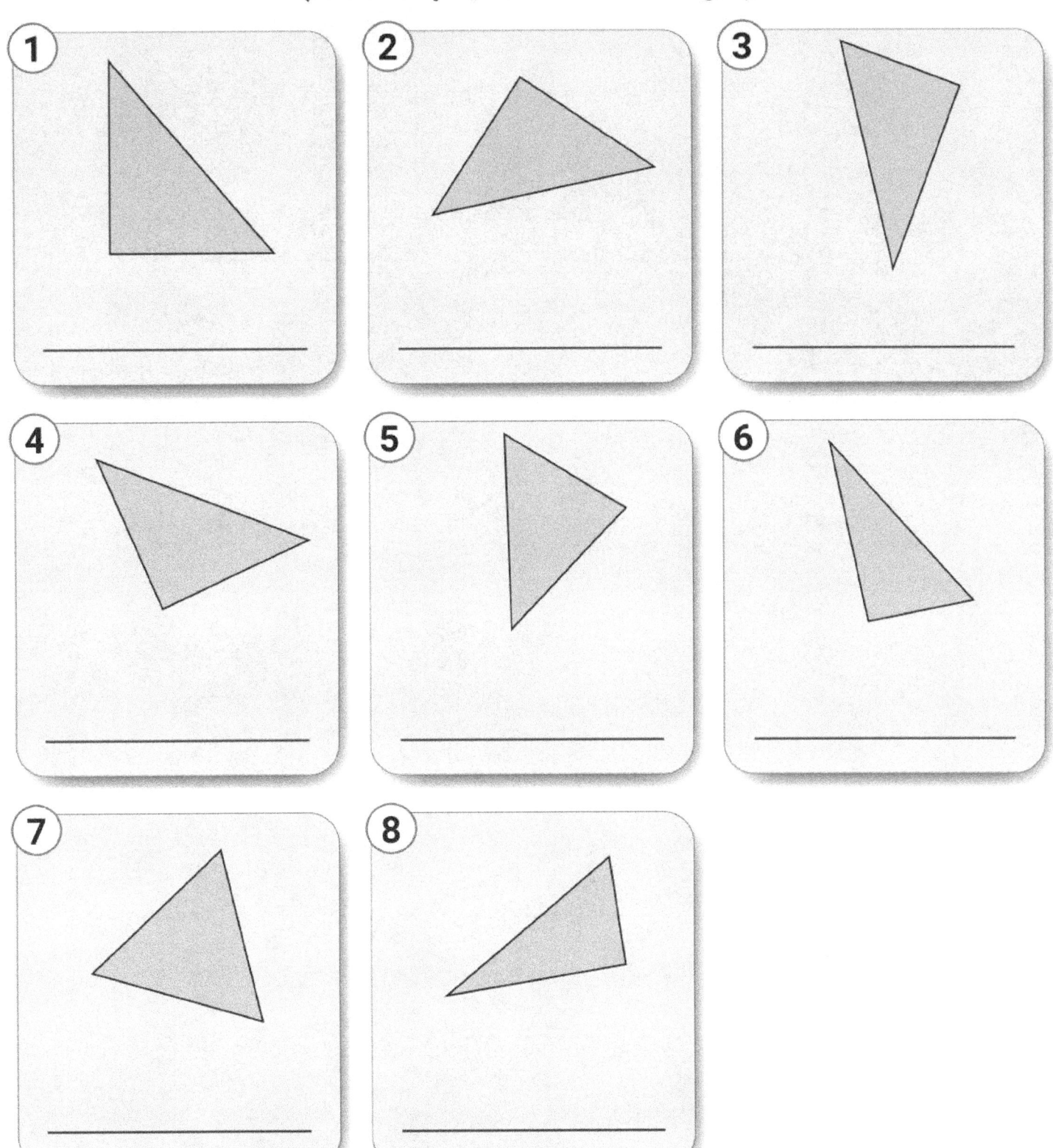

1.

2.

3.

4.

5.

6.

7.

8.

Classifying Triangles

Classify the following triangles by their sides and angles
(for example, isosceles and right).

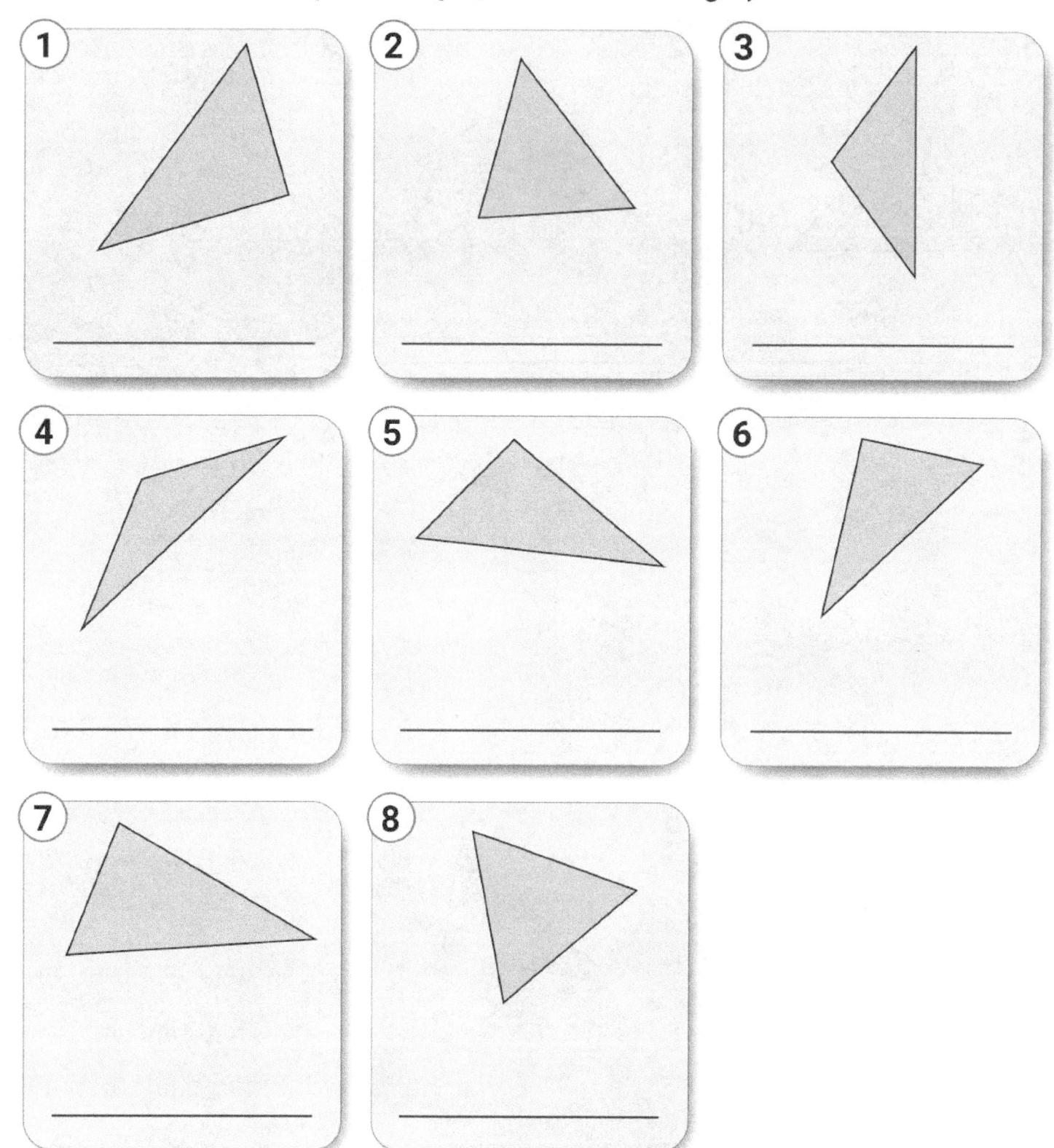

1.

2.

3.

4.

5.

6.

7.

8.

Calculating Perimeter

Determine the perimeter of the shapes.

1) A square with a side length of 8 inches.

2) A circle with a radius of 5 cm.

3) One triangle has side lengths of 4 inches, 9 inches, and 8 inches, while another equilateral triangle has a side length of 6 inches. Which triangle has a larger perimeter?

4) A hexagon with a side length of 7 cm.

5) A rectangular field is 37 meters wide and 84 meters long.

6) A rectangular bookmark that is 4 cm. wide and 12 cm. long.

7) A square with a side length of 15 mm.

8) A rhombus with a side length of 5 inches.

9) A trapezoid with side lengths of 19, 17, 15, and 17 yards.

10) A circle with a diameter of 14 feet.

Calculating Perimeter

Determine the perimeter of the shapes.

1. A regular hexagon with a side length of 12 cm.

2. A triangle with sides of 15, 17, and 18 inches.

3. A regular hexagon with a side length of 7 cm.

4. A regular octagon with a side length of 8 feet.

5. A regular pentagon with a side length of 3 yards.

6. A parallelogram with opposite sides of 13 and 19 inches.

7. A trapezoid with bases of 7 and 9 and sides of 4 inches.

8. A rectangle whose length is twice the width, which is 8 cm.

9. A triangle with two sides that are 14 inches and one side that is half as long.

10. A rectangle whose width is one-third the length, which is 27 cm.

Finding the Missing Side Lengths

Calculate the missing side lengths from the information provided.

1) The perimeter of a rectangular paper is 32 inches. One side is 7 inches.
   _____

2) The perimeter of a square piece of origami paper is 36 inches. What is the side length?
   _____

3) The perimeter of a square napkin is 20 inches. What is the side length?
   _____

4) The perimeter of a rectangular mirror is 58 inches. One side is 14 inches.
   _____

5) The perimeter of a rectangular placemat is 36 inches. One side is 10 inches.
   _____

6) The perimeter of a triangular main sail is 65 yards. One side is 29 yards and one is 19 yards.
   _____

7) The perimeter of a square playpen is 148 feet. What is the side length?
   _____

8) The perimeter of a triangle is 19 cm. One side is 7 and one is 8.
   _____

Finding the Missing Side Lengths

Calculate the missing side lengths from the information provided.

1. The perimeter of a regular hexagon is 48 centimeters. What is the length of a side?
   _____

2. A triangular chip has a perimeter of 17 centimeters. One side is 5 centimeters and another side is 6 centimeters.
   _____

3. A stop sign (an octagon) has a perimeter of 96 inches. What is the length of a side?
   _____

4. The perimeter of a rectangular pasture is 122 meters. One side is 35 meters.
   _____

5. The perimeter of a rectangular rug is 90 inches. One side is 18 inches.
   _____

6. The perimeter of a rectangular chalkboard is 140 inches. One side is 47 inches.
   _____

7. A regular pentagon has a perimeter of 45 centimeters. What is the side length?
   _____

8. The perimeter of a triangular bandana is 42 cm. One side is 17 cm and one is 13 cm.
   _____

Finding the Missing Side Lengths

Calculate the missing side lengths from the information provided.

1) The perimeter of a pentagon drawn on your paper is 85 millimeters. Four of the side lengths are as follows (in mm): 17, 11, 19, 18. What is the length of the other side?
___

2) The perimeter of a square cat litter box is 120 centimeters. What is the side length?
___

3) The perimeter of a triangle is 26 cm. One side is 9 and one is 12.
___

4) A six-sided polygon has a perimeter of 87 in. The sides (except 1) are as follows: 14, 13, 18, 22, 16.
___

5) A five-sided polygon has a perimeter of 45 in. The sides (except 1) are as follows: 4, 12, 9, 15.
___

6) An eight-sided polygon has a perimeter of 241 in. The sides (except 1) are as follows: 24, 39, 27, 31, 17, 33, 35.
___

7) A rectangular garden has an area of 88 square meters and one side is 8 meters. What is the length?
___

8) Grandma is making a quilt with square pieces. Each has an area of 81 square inches. What is the perimeter of each square?
___

Area of Triangles

Calculate the area of the following triangles.
Remember that the area of a triangle is calculated using the formula
$$A = \frac{1}{2}bh$$

1 a = 4 yards, b = 3 yards, c = 5 yards

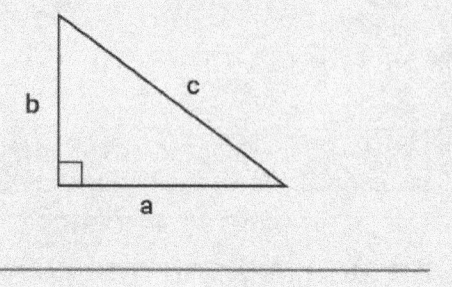

2 a = 6 cm, b = 7 cm, c = 8.5 cm, h = 3 cm

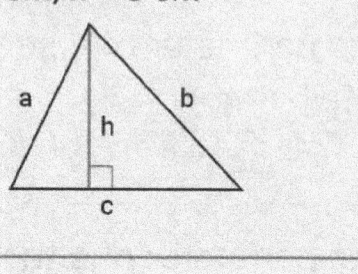

3 a = 52.72 feet, b = 90.07 feet, c = 98 feet, h = 48 feet

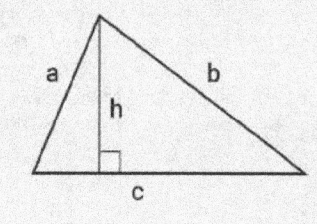

4 a = 5 feet, b = 12 feet, c = 13 feet

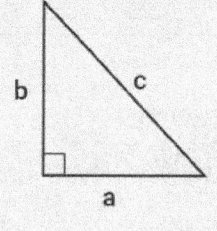

5 a = 41 meters, b = 65 meters

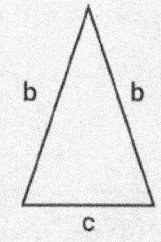

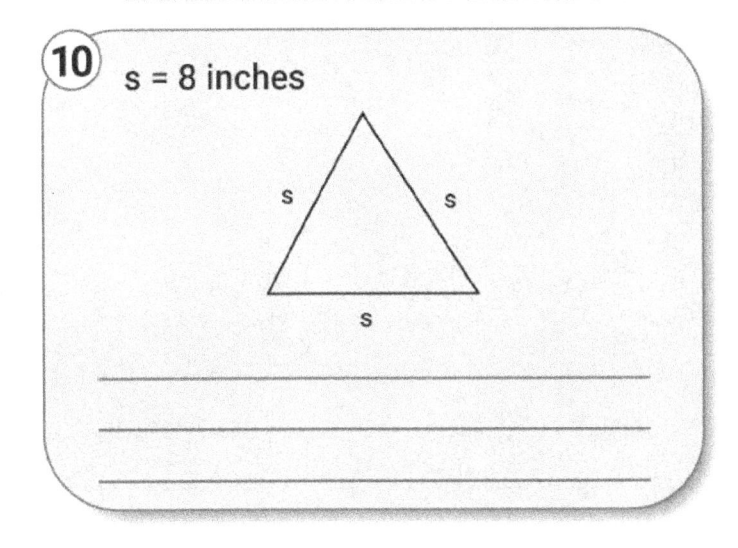

Area of Triangles

Calculate the area of the following triangles.
Remember that the area of a triangle is calculated using the formula
$$A = \frac{1}{2}bh$$

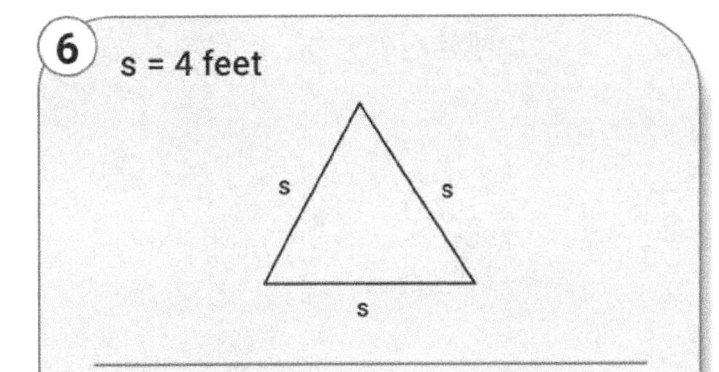

6) s = 4 feet

7) a = 24 cm, b = 7 cm, c = 25 cm

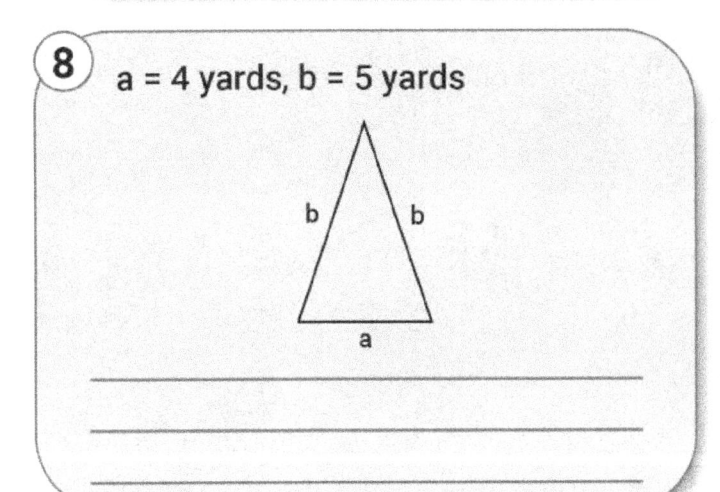

8) a = 4 yards, b = 5 yards

9) a = 16 mm, b = 12 mm, c = 20 mm

10) s = 8 inches

Area of Triangles

Calculate the area of the following triangles.
Remember that the area of a triangle is calculated using the formula
$$A = \frac{1}{2}bh$$

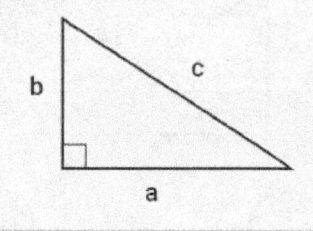

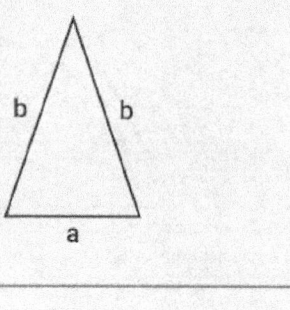

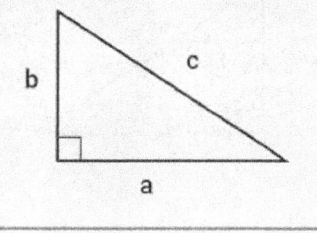

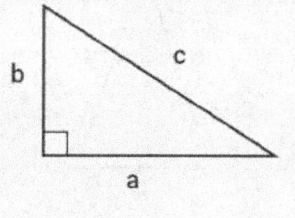

11) a = 84 mm, b = 56 mm, c = 100.96 mm

12) a = 10 feet, b = 12 feet

13) a = 70 yards, b = 56 yards, c = 89.64 yards

14) a = 75 cm, b = 55 mm, c = 93.01 mm

15) a = 72 inches, b = 58 inches

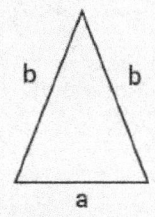

Circumference of Circles

Calculate the circumference of the following circles. Recall that the circumference of a circle is found by calculating πr². For the first column, report the exact measurement (using π). For the second column, use 3.14 for π. An example is shown below.

Example: A circle with a radius of 3 yards.

Exact Circumference	Approximate Circumference
9π yards	28.26 yards

1) A circle with a radius of 2 meters.
Exact Circumference _____   Approximate Circumference _____

2) A circle with a radius of 1 inch.
Exact Circumference _____   Approximate Circumference _____

3) A circle with a diameter of 12 feet.
Exact Circumference _____   Approximate Circumference _____

4) A circle with a diameter of 26 mm.
Exact Circumference _____   Approximate Circumference _____

5) A circle with a diameter of 18 feet.
Exact Circumference _____   Approximate Circumference _____

6) A circle with a radius of 10 meters.
Exact Circumference _____   Approximate Circumference _____

7) A circle with a radius of 5 centimeters.
Exact Circumference _____   Approximate Circumference _____

8) A circle with a radius of 8 inches.
Exact Circumference _____   Approximate Circumference _____

9) A circle with a diameter of 14 inches.
Exact Circumference _____   Approximate Circumference _____

10) A circle with a diameter of 24 feet.
Exact Circumference _____   Approximate Circumference _____

Area of Mixed Shapes and Figures

Determine the area of the following shapes and figures and round to the nearest tenth of a decimal.

1) A square with a side length of 3 meters.

2) A rectangle that is 11 feet by 5 feet.

3) A rectangle that is 8 inches by 7 inches.

4) A square with a side length of 6 centimeters.

5) A triangle with a base of 4 inches and a height of 7 inches.

6) A triangle with a base of 12 cm and a height of 8 cm.

7) A square with a side length of 15 inches.

8) A rectangle with a length of 12 inches and a width of 5 inches.

9) A circle with a radius of 9 cm.

10) A triangle with a base of 4 inches and a height of 5 inches.

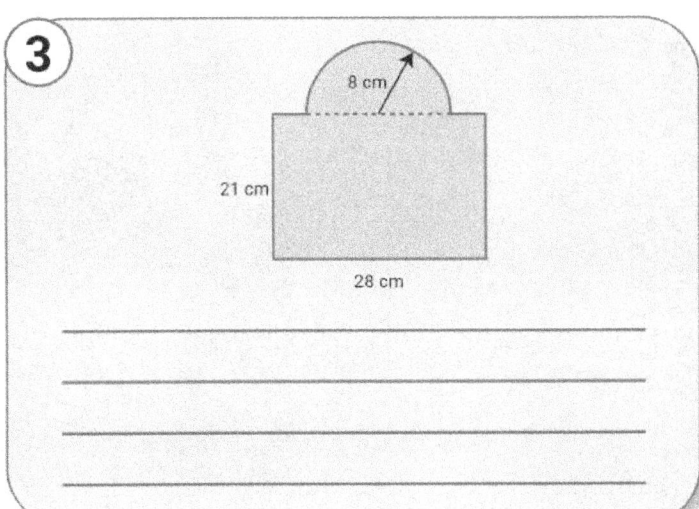

Area of Mixed Shapes and Figures

Determine the area of the following shapes and figures

1 10 yd, 10 yd, 30 yd

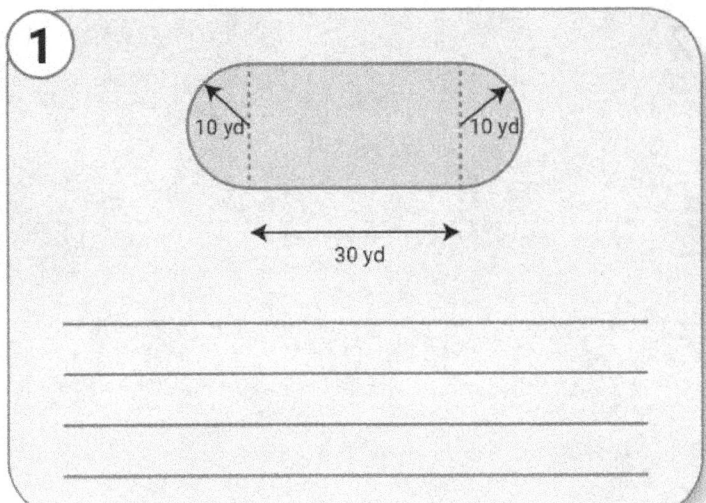

2 3 in, 2 in

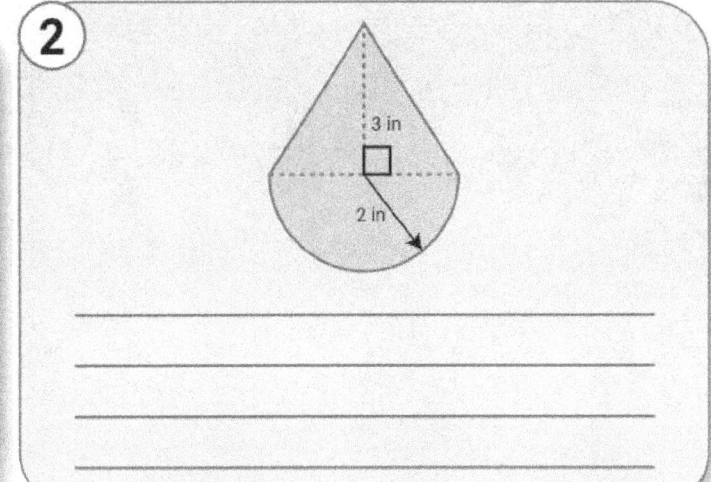

3 8 cm, 21 cm, 28 cm

4 9 ft, 9 ft, 4 ft

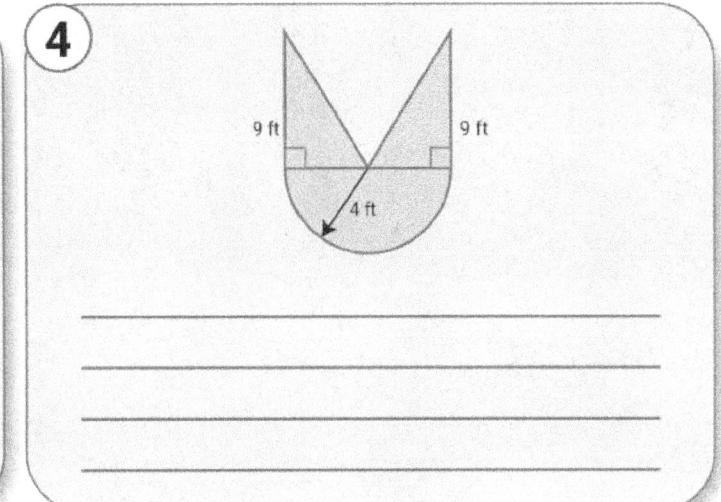

5 9 m, 27 m

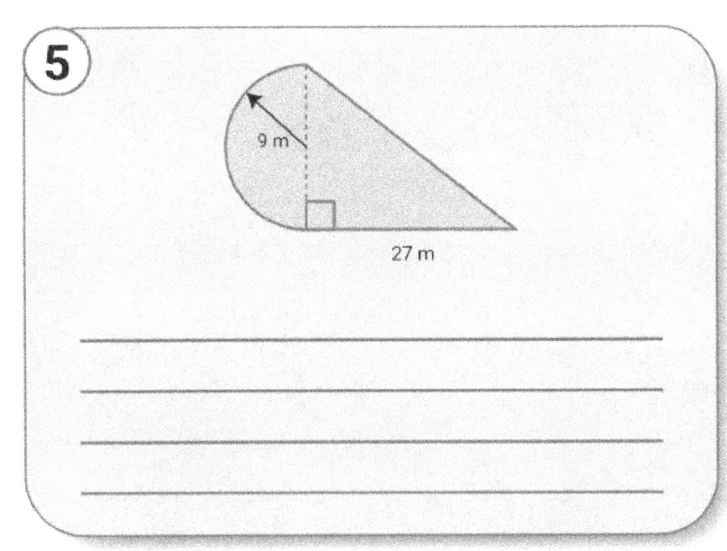

6 14 m, 12 m, 15 m, 12 m

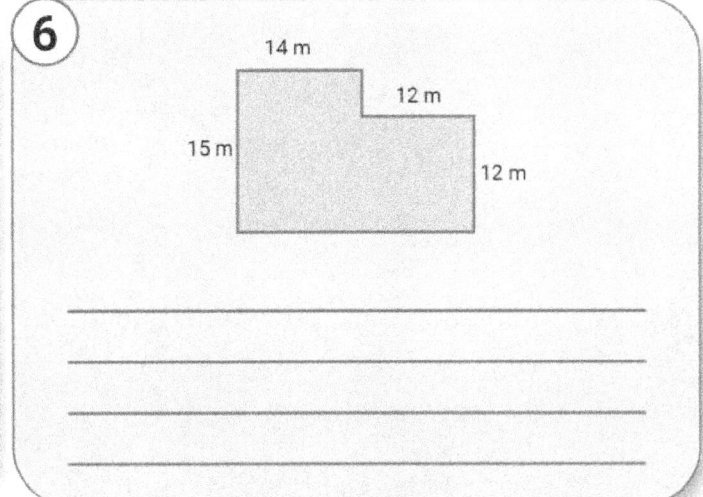

Area of Mixed Shapes and Figures

Determine the area of the following shapes and figures

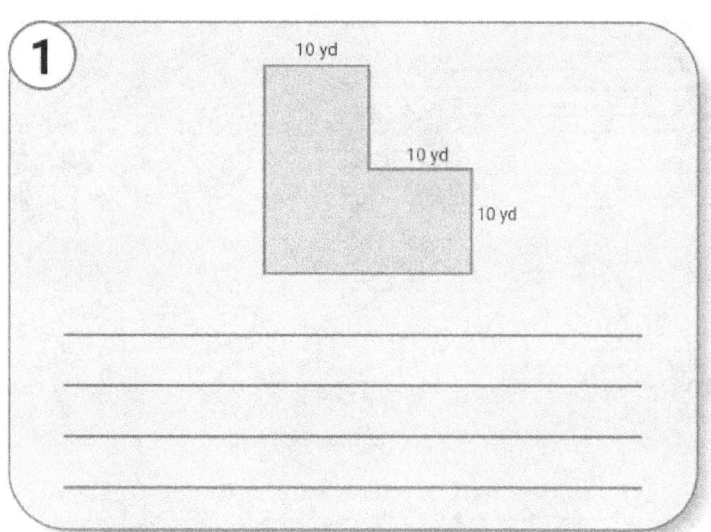

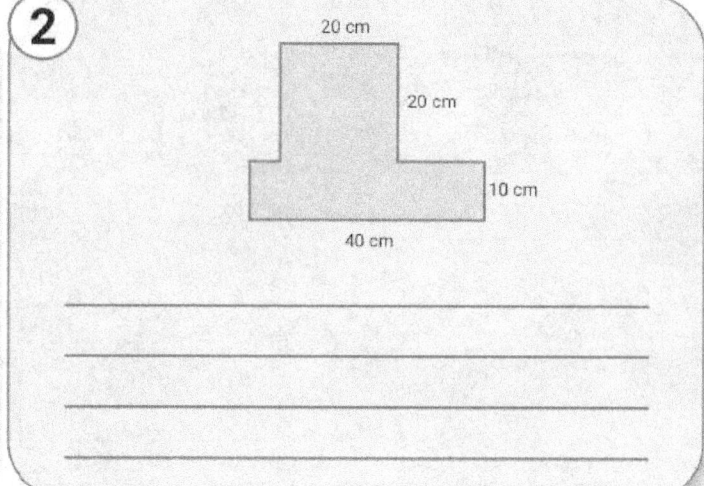

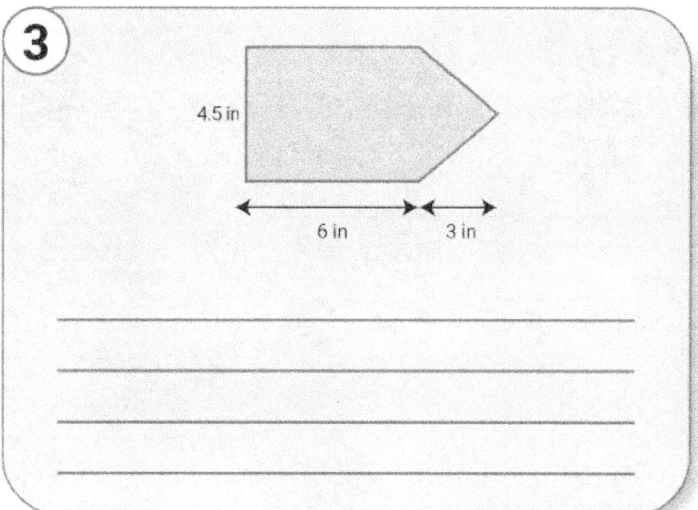

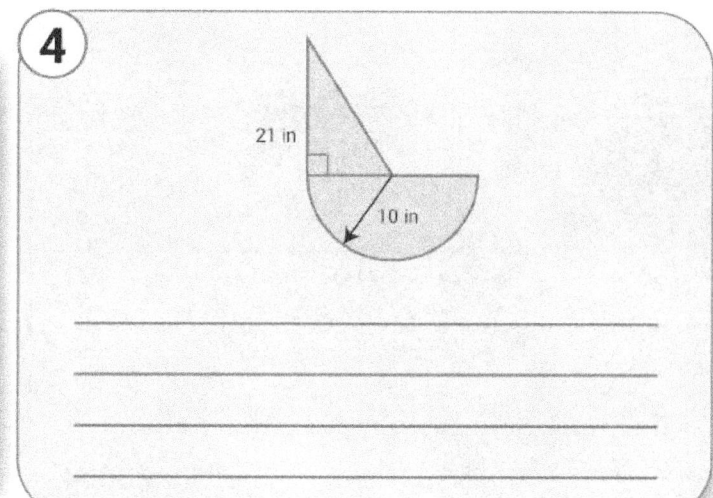

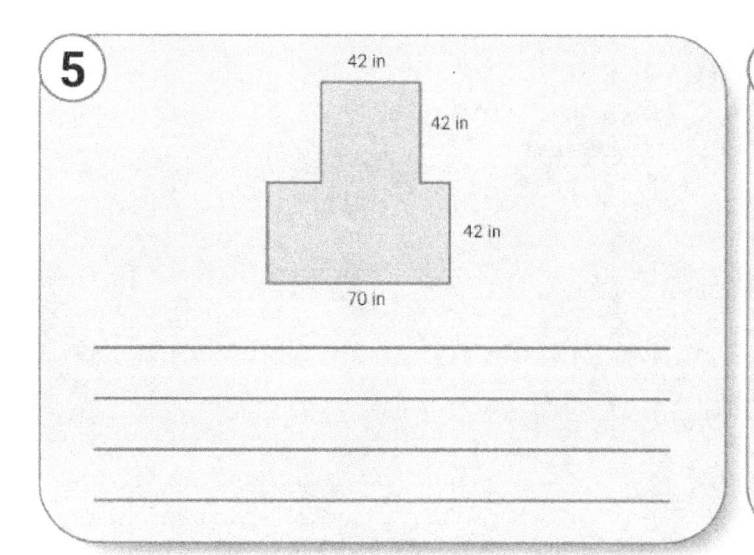

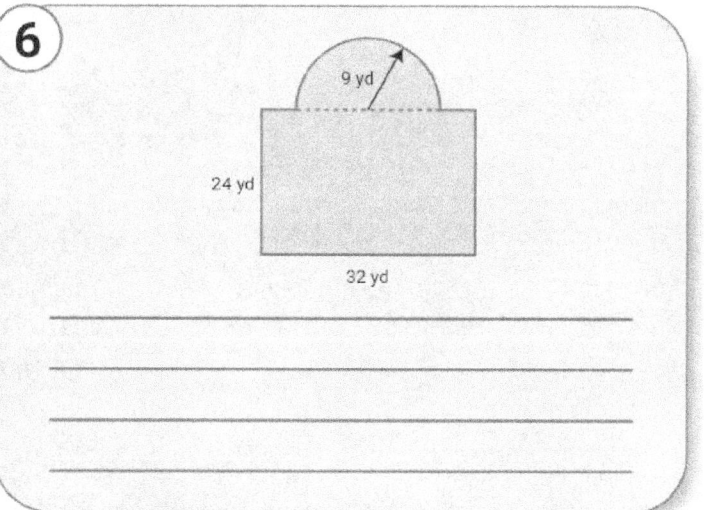

1
10 yd
10 yd
10 yd

2
20 cm
20 cm
10 cm
40 cm

3
4.5 in
6 in   3 in

4
21 in
10 in

5
42 in
42 in
42 in
70 in

6
9 yd
24 yd
32 yd

Volume

Find the volumes of the following solids. Use 3.14 for π where necessary.

1) A small cube-shaped tissue box that has a side length of 12 cm.

2) A ring box with a side length of 18 mm.

3) A cereal box that is 3 inches wide, 7 inches wide, and 11 inches tall.

4) A kid's soccer ball with a diameter of 8 inches.

5) A rectangular swimming pool that is 25 yards long, 12 yards wide, and 2 yards deep.

6) The Great Pyramid with a height of 146 m and a side of the square base of 230 m.

7) A snow cone with a length (height) of 5 inches and a radius of 2 inches.

8) A chocolate box with a length of 12 inches, a width of 8 inches, and a depth of 3 inches.

Volume

Find the volumes of the following solids. Use 3.14 for π where necessary.

1. A sand castle in the shape of a rectangular pyramid with a base side lengths of 5 inches and 9 inches and a height of 10 inches.

2. A cylindrical water tank that is 80 meters high and has a radius of 12 meters.

3. A shipping box that is 15 inches by 7 inches by 8 inches.

4. A fish tank that is 24 inches by 18 inches by 16 inches.

5. A play teepee in the shape of a cone that is 6 feet high and has a diameter of 8 feet.

6. A farmer's grain storage container (cylindrical) that is 8 meters high and has a diameter of 4 meters.

7. A jewelry box with a length of 10 inches, a depth of 5 inches, and a width of 7 inches.

8. A can of soup that is 4 inches tall and has a radius of 2 inches.

Volume

Find the volumes of the following solids. Use 3.14 for π where necessary.

1. The sphere of a gumball machine that has a radius of 5 inches.

2. A little bouncy rubber ball with a diameter of 6 cm.

3. A traffic cone with a height of 36 inches and a diameter of 12 inches.

4. A large gourmet spherical lollipop with a radius of 1 inch.

5. A boy has a toy chest that is 3 feet long, 2 feet tall, and 2 feet deep. What is the volume of the toy chest?

6. A girl is baking a cake. She bakes it in a 9-inch by 13-inch pan. Once the cake rises it is 3 inches tall. What is the total volume of the cake she will frost if she does the top and all four sides?

7. A snare drum has a radius of 6 inches. It has a height of 8 inches. What is the volume of the drum? Use 3.14 for π.

8. Shotaro is stacking hay bales. Each hay bale is 36 inches by 30 inches by 40 inches. If he has 12 rows of hay bales going across and each row has 6 hay bales piled high, what total volume is occupied by the hay bales?

Are They Similar?

Determine if the following polygons are similar.
Circle the correct classification for each pair.

1

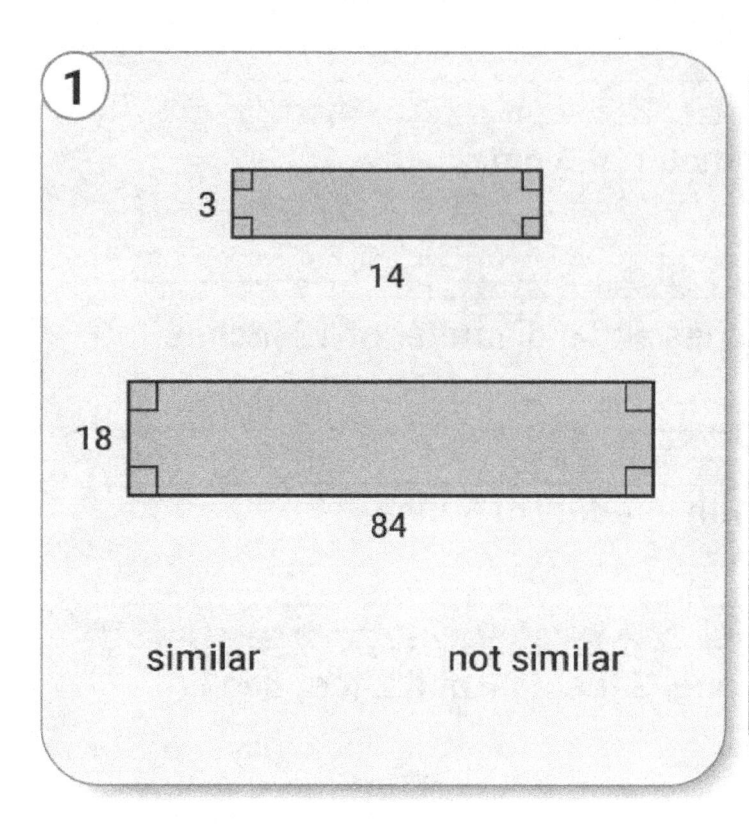

3, 14
18, 84

similar   not similar

2

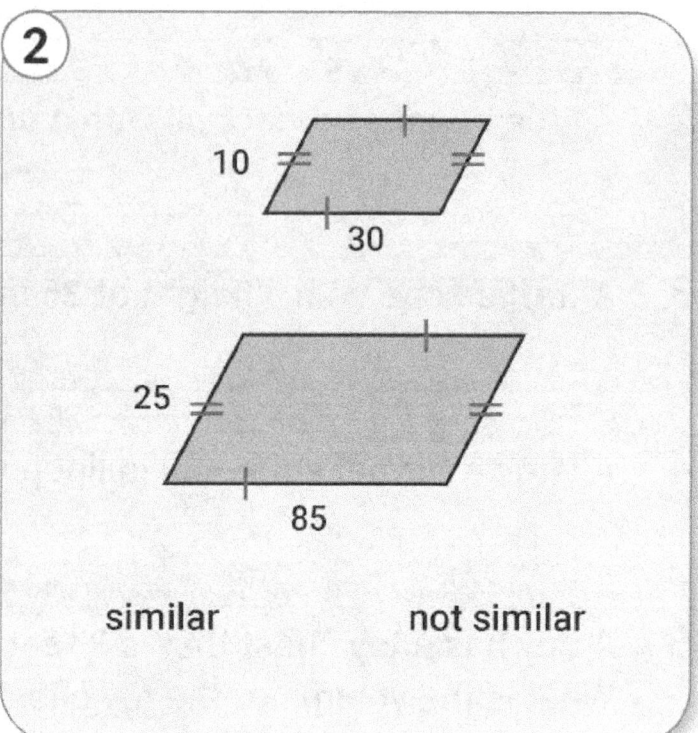

10, 30
25, 85

similar   not similar

3

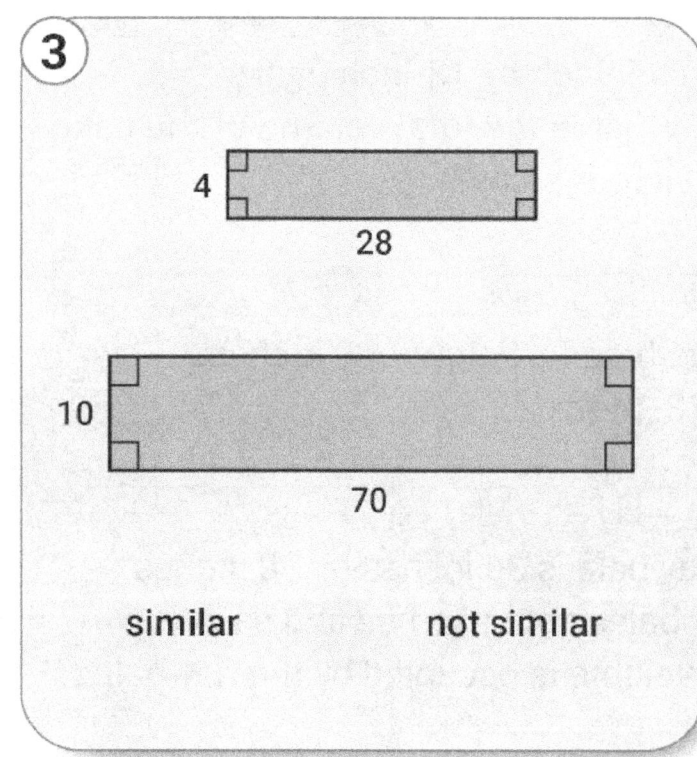

4, 28
10, 70

similar   not similar

4

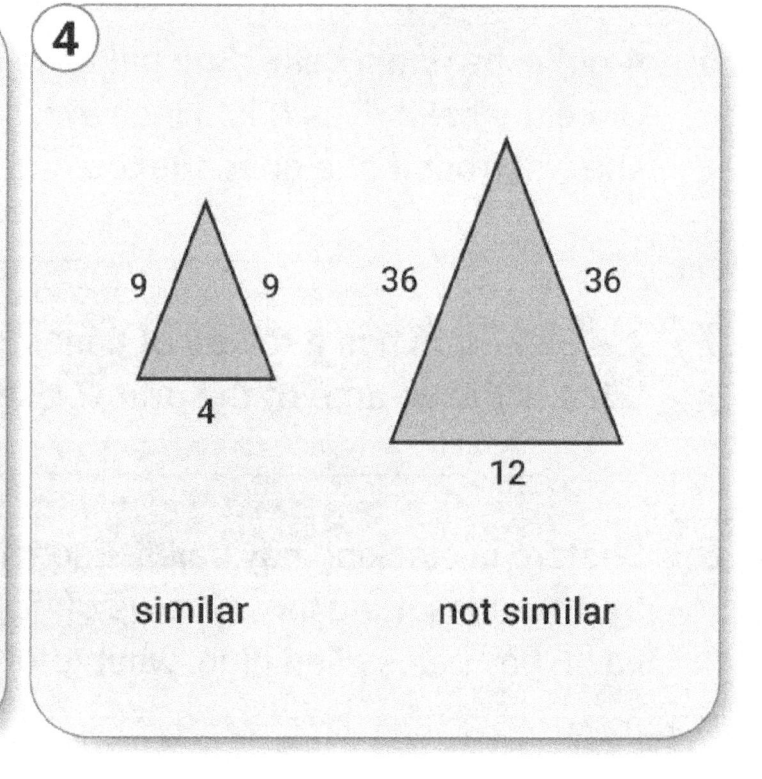

9, 9, 4
36, 36, 12

similar   not similar

Are They Similar?

Determine if the following polygons are similar.
Circle the correct classification for each pair.

1

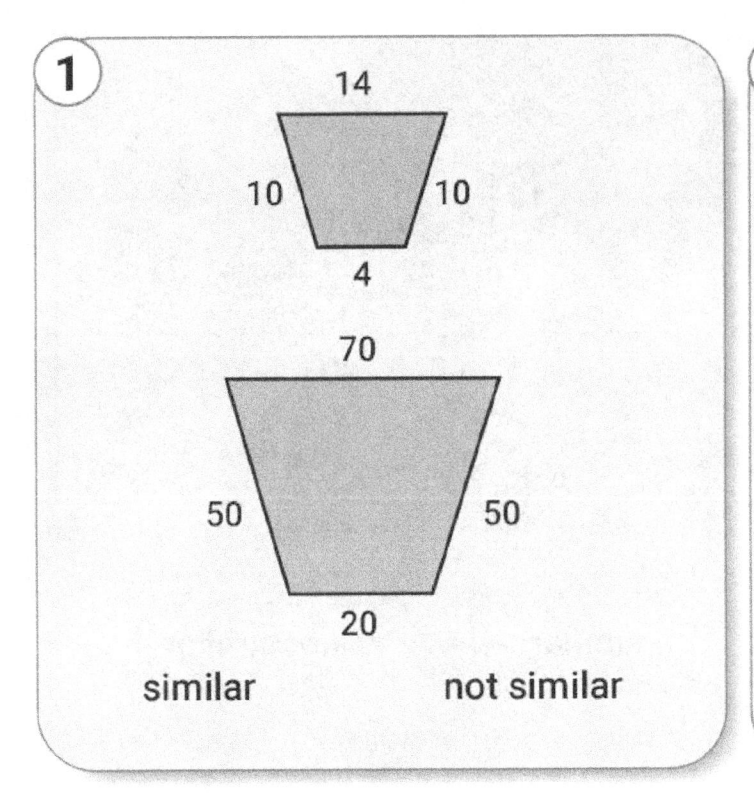

Trapezoid: 14 (top), 10 (left), 10 (right), 4 (bottom)
Trapezoid: 70 (top), 50 (left), 50 (right), 20 (bottom)

similar   not similar

2

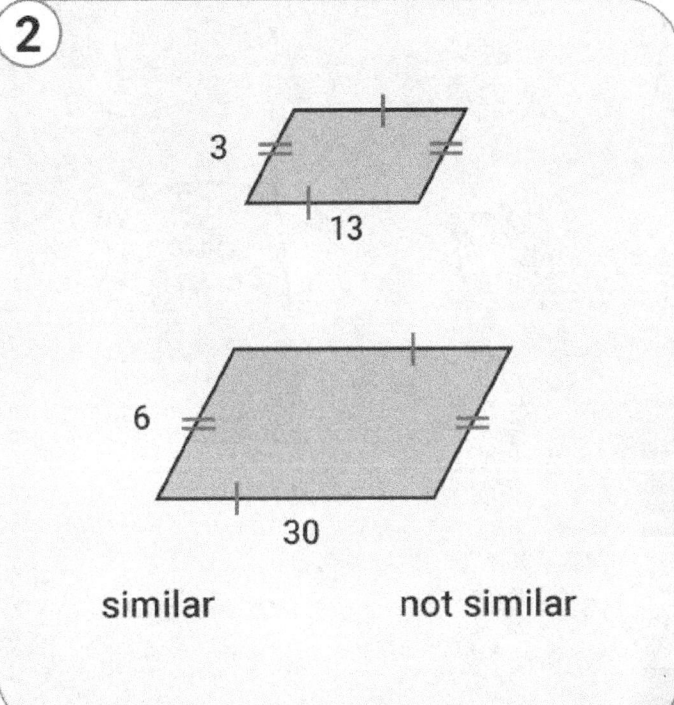

Parallelogram: 3, 13
Parallelogram: 6, 30

similar   not similar

3

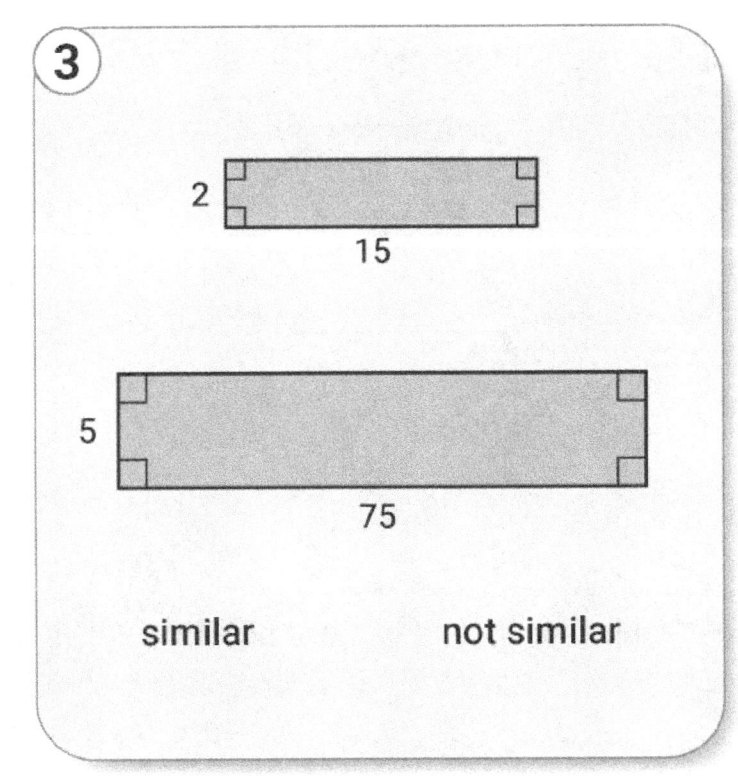

Rectangle: 2 by 15
Rectangle: 5 by 75

similar   not similar

4

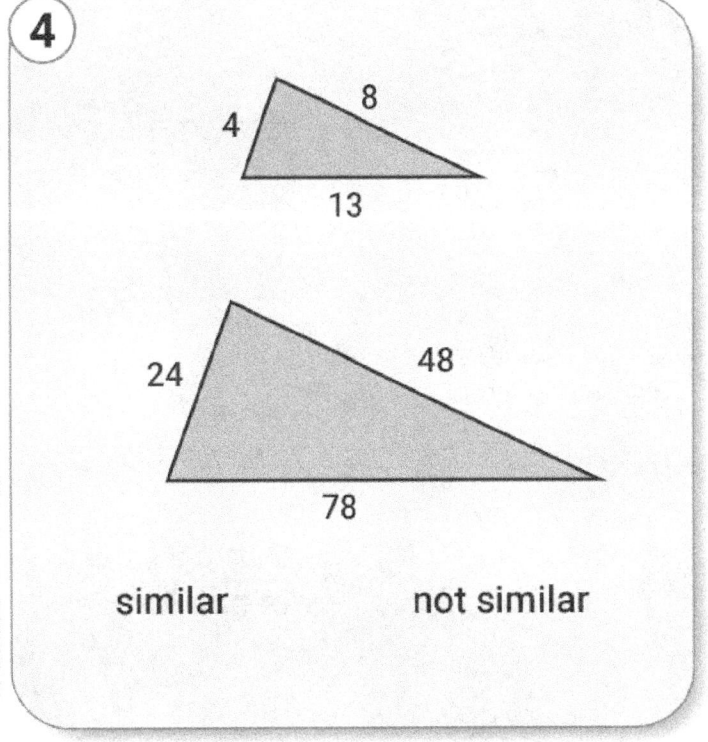

Triangle: 4, 8, 13
Triangle: 24, 48, 78

similar   not similar

Are They Similar?

Determine if the following polygons are similar.
Circle the correct classification for each pair.

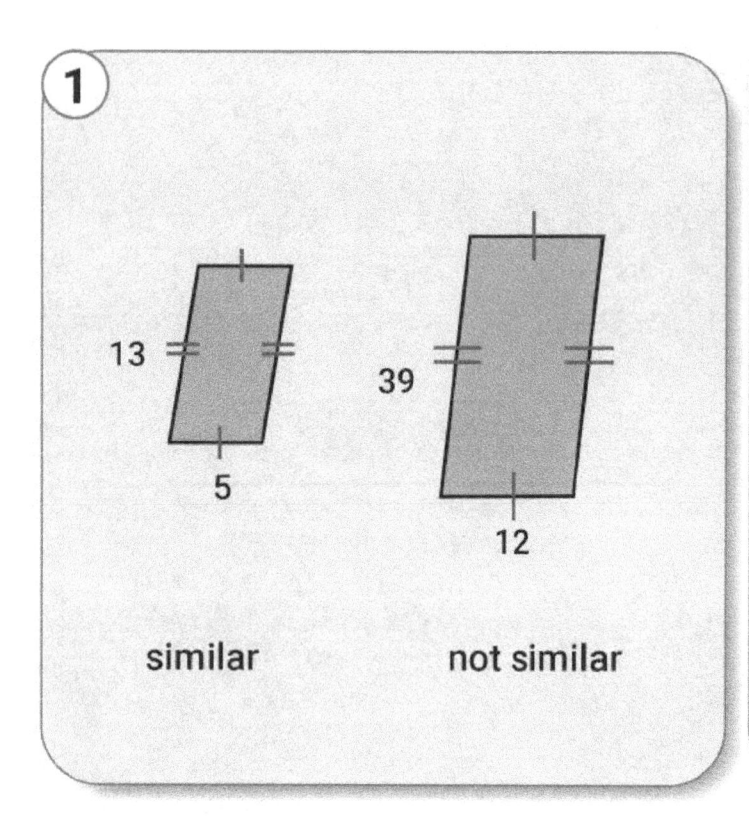

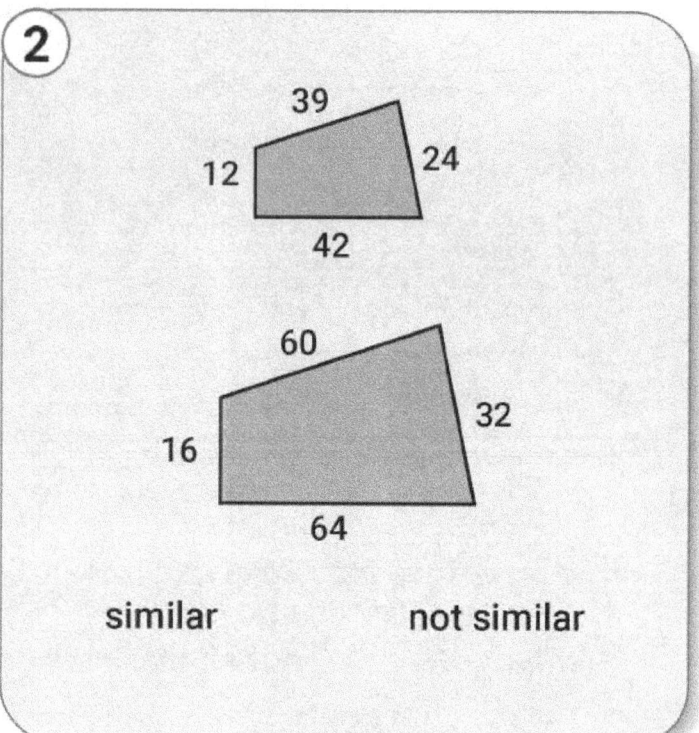

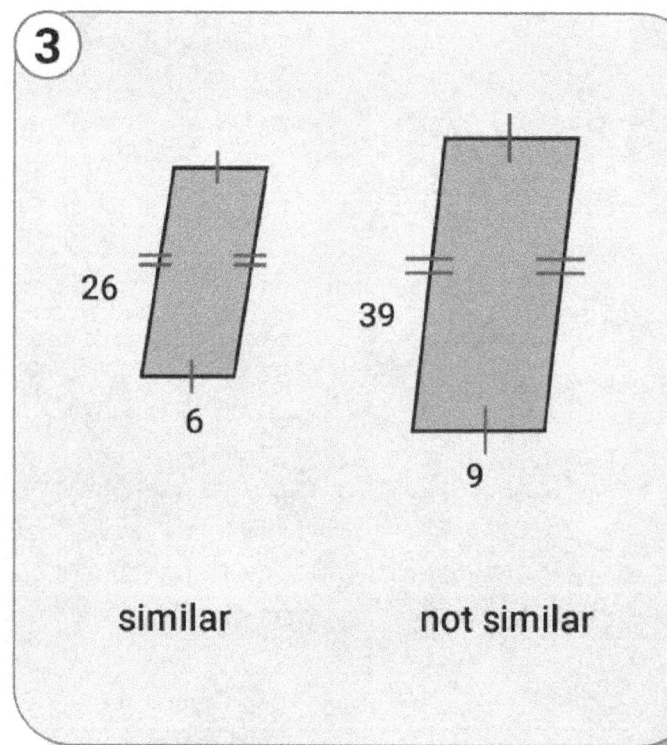

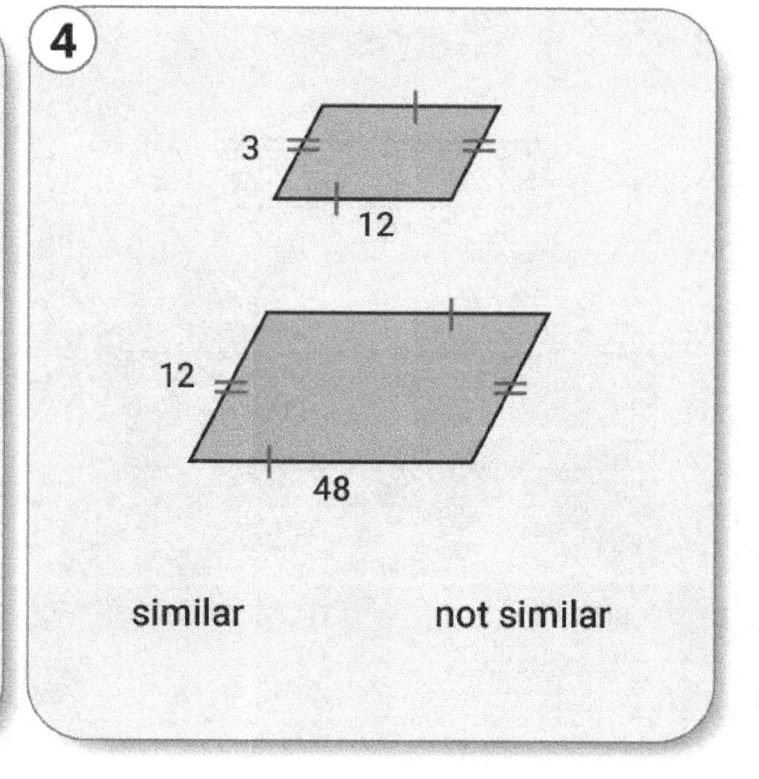

1)

similar   not similar

2)

similar   not similar

3)

similar   not similar

4)

similar   not similar

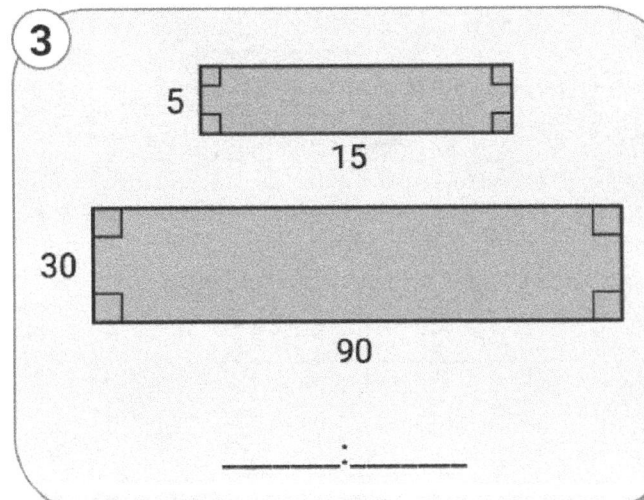

Find the Scale Factor

Each of the following polygon pairs is similar. Determine the scale factor of the smaller figure to the larger figure.

1)
Small rectangle: 2 by 12
Large rectangle: 4 by 24

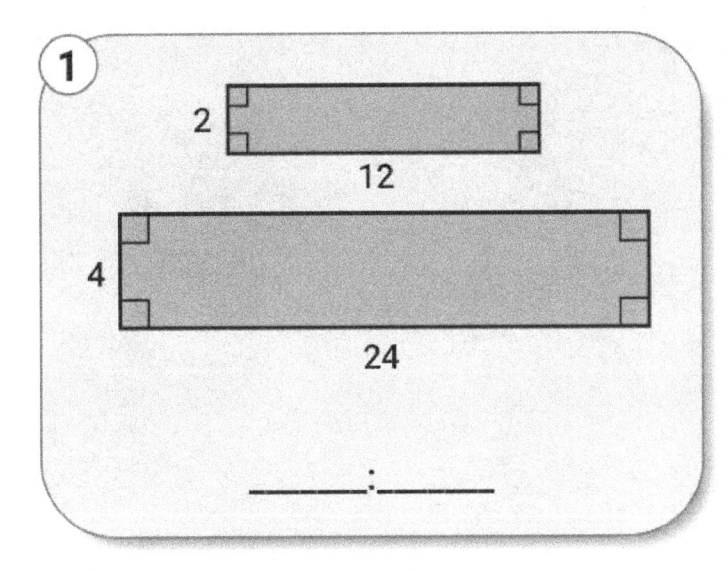

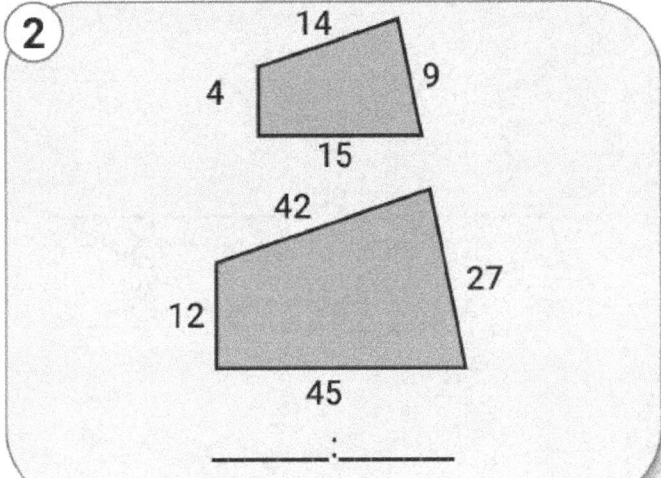

_____ : _____

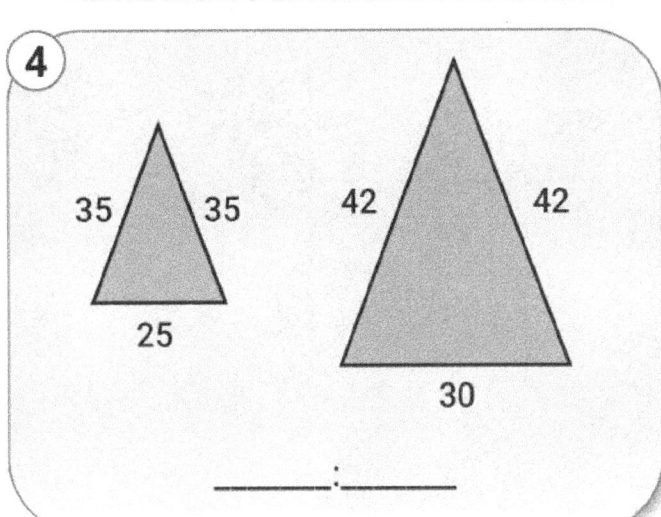

2)
Small quadrilateral: 14, 9, 15, 4
Large quadrilateral: 42, 27, 45, 12

_____ : _____

3)
Small rectangle: 5 by 15
Large rectangle: 30 by 90

_____ : _____

4)
Small triangle: 35, 35, 25
Large triangle: 42, 42, 30

_____ : _____

5)
Small rectangle: 9 by 36
Large rectangle: 15 by 60

_____ : _____

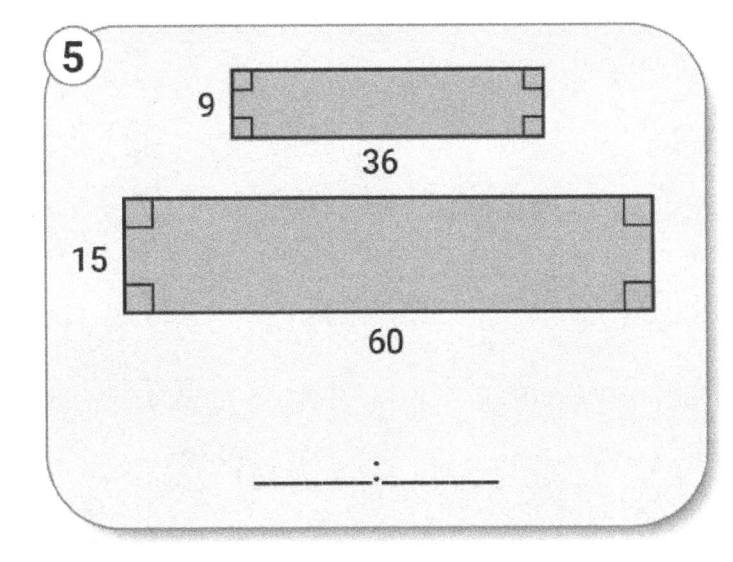

Find the Scale Factor

Each of the following polygon pairs is similar. Determine the scale factor of the smaller figure to the larger figure.

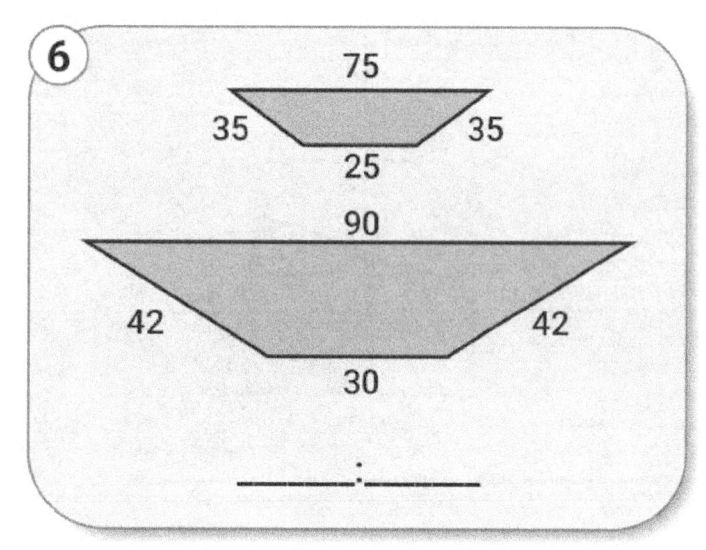

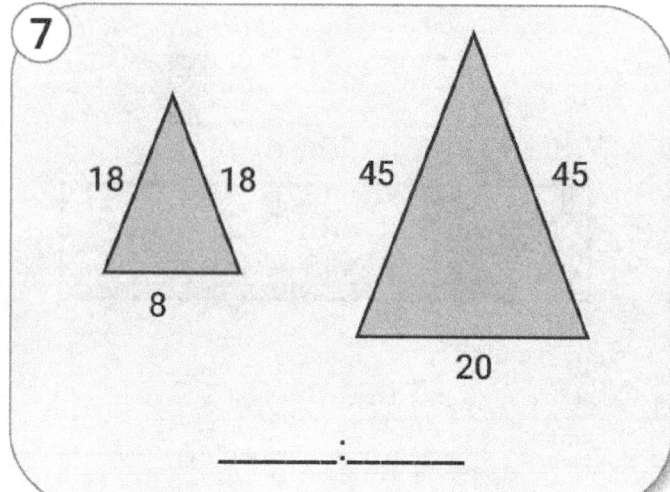

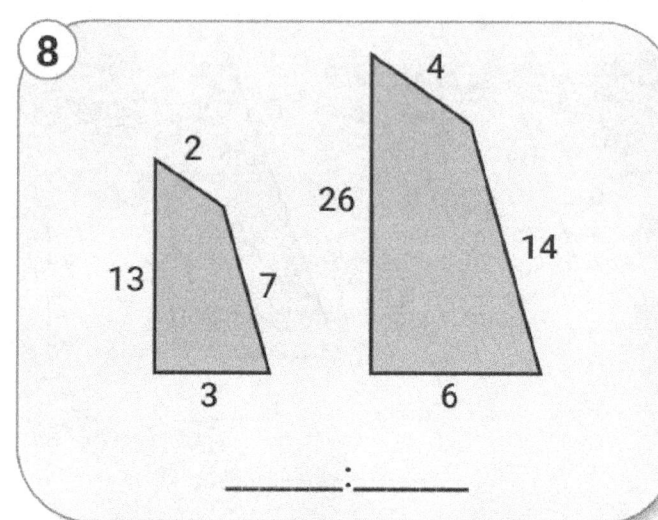

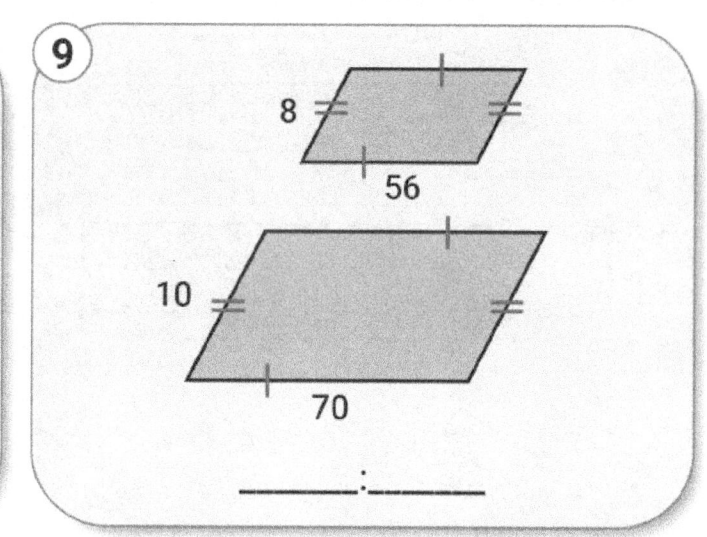

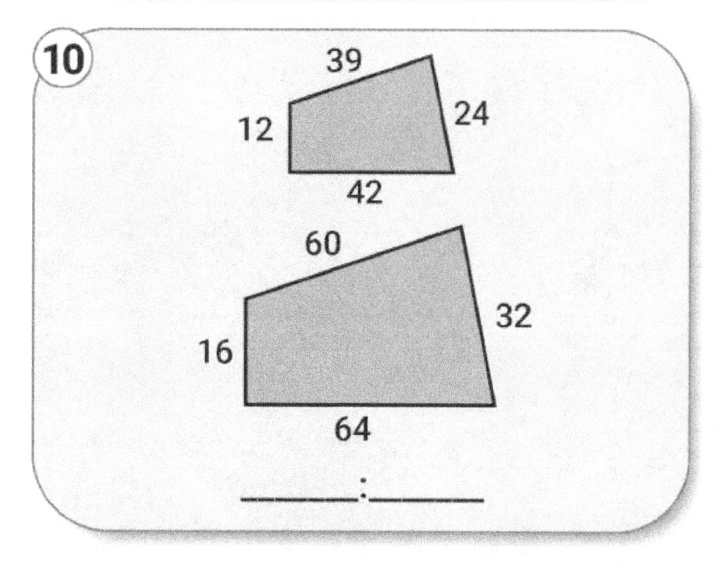

6
- Smaller trapezoid: 75 (top), 35 (sides), 25 (height?)
- Larger trapezoid: 90 (top), 42 (sides), 30 (bottom)

_____ : _____

7
- Smaller triangle: 18, 18, 8
- Larger triangle: 45, 45, 20

_____ : _____

8
- Smaller figure: 2, 13, 7, 3
- Larger figure: 4, 26, 14, 6

_____ : _____

9
- Smaller parallelogram: 8, 56
- Larger parallelogram: 10, 70

_____ : _____

10
- Smaller quadrilateral: 39, 12, 24, 42
- Larger quadrilateral: 60, 16, 32, 64

_____ : _____

The Pythagorean Theorem

For each of the following right triangles described, circle the measure that would be the closest estimate of the hypotenuse.

1) Side lengths of 6 and 7 inches.

  a) 5 inches   b) 8 inches   c) 9 inches

2) Side lengths of 10 and 10 inches.

  a) 13 inches   b) 14 inches   c) 15 inches

3) Side lengths of 6 and 4 inches.

  a) 7 inches   b) 8 inches   c) 9 inches

4) Side lengths of 5 and 8 inches.

  a) 7 inches   b) 9 inches   c) 11 inches

5) Side lengths of 8 and 12 inches.

  a) 10 inches   b) 13 inches   c) 14 inches

Independent Probability

1) In a standard deck of 52 playing cards, what's the probability of drawing a heart?
_____

2) When you roll a single playing die, what's the probability you roll a number less than a 3?
_____

3) In a standard deck of 52 playing cards, what's the probability of drawing a face card?
_____

4) In a standard deck of 52 playing cards, what's the probability of drawing a 3?
_____

5) When you roll a single playing die, what's the probability you roll a number that is a divisor of 8?
_____

6) When you roll a single playing die, what's the probability you roll a number that is a factor of 12?
_____

7) When you roll a pair of dice, what's the total number of possible outcomes rolled?
_____

8) In a standard deck of 52 playing cards, what's the probability of drawing a red card on the first draw, replacing it, and drawing a red card on the second draw?
_____

9) When you roll a single playing die, what's the probability you roll a number that is a factor of 20?
_____

10) If you roll a pair of dice, what's the probability that you roll a 3 or less on the first die and a 2 or less on the second die?
_____

Dependent Probability

1) A bag contains 6 lemon drops and 4 cherry drops. What is the probability that if one candy is taken out and it is a lemon drop, and then a second candy is taken out without putting the first one back that it is also a lemon drop?

2) Jeff is selecting names of teammates for his debate team out of hat. If there are 4 boys and 5 girls to choose from, what is the probability he will select two boys' names if he does not put the names back in the hat after drawing them.

3) A bag contains 9 red marbles, 5 blue marbles, 8 yellow marbles, and 3 white marbles. If three marbles are drawn and not replaced after drawing them, what is the probability that all three are red?

4) Candice has a bag of toys to give out to hospital patients. The bag contains 5 trucks, 4 soccer balls, 1 teddy bear, and 3 dolls. What is the probability of selecting a teddy bear and then a soccer ball?

5) There are 8 cars and 5 trucks. What is the probability of drawing 5 trucks with 5 draws?

Mean, Median, Mode, and Range

Determine the mean, median, mode, and range for each of the provided data sets. Round each answer to the nearest whole number.

① 12, 68, 20, 32, 73

Mean: ____, Median: ____,

Mode: ____, Range: ____

② 69, 87, 55, 42, 25, 32, 69, 45

Mean: ____, Median: ____,

Mode: ____, Range: ____

③ 89, 38, 97, 79, 73, 66, 55

Mean: ____, Median: ____,

Mode: ____, Range: ____

④ 65, 63, 72, 68, 61, 75, 23

Mean: ____, Median: ____,

Mode: ____, Range: ____

⑤ 52, 76, 84, 80, 92, 62, 48, 74

Mean: ____, Median: ____,

Mode: ____, Range: ____

⑥ 40, 16, 44, 39, 59, 22, 29, 58, 44

Mean: ____, Median: ____,

Mode: ____, Range: ____

⑦ 41, 31, 37, 16, 62, 20, 53, 37, 83, 60

Mean: ____, Median: ____,

Mode: ____, Range: ____

⑧ -16, 79, -58, -57, -75, 95, 27, 51

Mean: ____, Median: ____,

Mode: ____, Range: ____

⑨ 51, 77, -81, 36, 27, 71, -76

Mean: ____, Median: ____,

Mode: ____, Range: ____

⑩ -71, 76, 82, 80, 83

Mean: ____, Median: ____,

Mode: ____, Range: ____

Patterns of Association in Bivariate Data

For each of the following pairs of variables, predict what kind of association, if any, exists for the pair.

Variable Pair	Association
Person's age and eyesight ability	
Hours spent studying and exam grade	
Temperature and amount of clothing worn	
Income and calories consumed	
Population density and crime rate	
Amount of soda consumed and number of cavities	
Car insurance cost and number of years without a traffic violation	
Age and grade point average	
Customer satisfaction and number of defective products	
Shoe size and mile running time	

Answers

Prime or Composite?

1. Prime	2. Prime	3. Composite
4. Composite	5. Prime	6. Prime
7. Composite	8. Prime	9. Prime
10. Composite	11. Composite	12. Composite
13. Prime	14. Composite	15. Prime

Greatest Common Factor

1) 3	2) 1	3) 15	4) 8	5) 6
6) 2	7) 4	8) 24	9) 1	10) 14
11) 18	12) 7	13) 8	14) 29	15) 6

Least Common Multiple

1) 30	2) 60	3) 60	4) 120	5) 20
6) 30	7) 120	8) 24	9) 120	10) 30
11) 120	12) 121	13) 18	14) 91	15) 60
16) 120				

Prime Factorization

1) 115 = 5 x 23
2) 136 = 2 x 2 x 2 x 17
3) 45 = 3 x 3 x 5
4) 104 = 2 x 2 x 2 x 13
5) 28 = 2 x 2 x 7
6) 60 = 2 x 2 x 3 x 5
7) 114 = 2 x 3 x 19
8) 88 = 2 x 2 x 2 x 11
9) 98 = 2 x 7 x 7
10) 96 = 2 x 2 x 2 x 2 x 2 x 3
11) 81 = 3 x 3 x 3 x 3
12) 110 = 2 x 5 x 11
13) 63 = 3 x 3 x 7
14) 130 = 2 x 5 x 13
15) 42 = 2 x 3 x 7

Order of Operations

1)
$(10 - 3) \times (9 - 6) + 7^2$
$= (10 - 3) \times (9 - 6) + 49$
$= 7 \times 3 + 49$
$= 21 + 49$
$= 70$

2)
$(15 - 7) \times (12 - 6) + 6^2$
$= (15 - 7) \times (12 - 6) + 36$
$= 8 \times 6 + 36$
$= 48 + 36$
$= 84$

3)
$(10 - 3)^2 + (12 - 15 \div 5)$
$= 49 + (12 - 3)$
$= 49 + 9$
$= 58$

4)
$2 \times (6 \times 3 - 8^2) + 22$
$= 2 \times (6 \times 3 - 8^2) + 22$
$= 2 \times (6 \times 3 - 64) + 22$
$= 2 \times (18 - 64) + 22$
$= 2 \times -46 + 22$
$= -92 + 22$
$= -70$

5)
$(9 + 56 - 5) \div 2 + 3^2$
$= (9 + 56 - 5) \div 2 + 3^2$
$= 60 \div 2 + 3^2$
$= 60 \div 2 + 9$
$= 30 + 9$
$= 39$

6)
$(11 + 53 - 4^2) \div (9 + 7)$
$= (11 + 53 - 4^2) \div (9 + 7)$
$= (11 + 53 - 16) \div (9 + 7)$
$= (64 - 16) \div (16)$
$= 48 \div 16$
$= 3$

7)
$(14 + 19 - 3^2) \div (12 \div 2)$
$= (14 + 19 - 9) \div (12 \div 2)$
$= (33 - 9) \div 6$
$= 24 \div 6$
$= 4$

8)
$(3^2 - 4)^2 + (16 + 20 \div 10)$
$= (9-4)^2 + (16 + 2)$
$(5)^2 + 18$
$= 25 + 18$
$= 43$

9)
$3(8 \div 2^2)^3 + (5 \times 8)$
$= 3(8 \div 4)^3 + 40$
$= 3(2)^3 + 40$
$= 3(8) + 40$
$= 24 + 40$
$= 64$

10)
$(2 \times 6^2 + 8) \div (8 - 20 \div 5)^2$
$= (2 \times 36 + 8) \div (8 - 4)^2$
$= 80 \div 16$
$= 5$

Order of Operations

11) 13: $4 + (3 \times 2)^2 \div 4$
$= 4 + 6^2 \div 4$
$= 4 + 36 \div 4$
$= 4 + 9$
$= 13$

12) 2: $2 \times (6 + 3) \div (2 + 1)^2$
$= 2 \times 9 \div (3)^2$
$= 2 \times 9 \div 9$
$= 18 \div 9$
$= 2$

13) 7: $2^2 \times (3 - 1) \div 2 + 3$
$= 2^2 \times 2 \div 2 + 3$
$= 4 \times 2 \div 2 + 3$
$= 8 \div 2 + 3$
$= 4 + 3$
$= 7$

14) 65: $(12 + 3) \times (8 - 2) - 5^2$
$= 15 \times 6 - 5^2$
$= 15 \times 6 - 25$
$= 90 - 25$
$= 65$

15) 114: $(19 - 8) \times (13 - 3) + 2^2$
$= 11 \times 10 + 2^2$
$= 11 \times 10 + 4$
$= 110 + 4$
$= 114$

16) 48: $(2 + 4)^2 + (9 + 12 \div 4)$
$= (6)^2 + (9 + 3)$
$= 36 + 12$
$= 48$

17) 297: $3 \times (13 \times 3 + 8^2) - 12$
$3 \times (13 \times 3 + 64) - 12$
$= 3 \times (103) - 12$
$= 309 - 12$
$= 297$

18) 63: $[(4+3)^2 + 1] + 2^3 - 5$
$= (7 + 11) + 2^3 - 5$
$= 60 + 8 - 5$
$= 63$

19) 8: $[6^2 + (20 \div 5 + 4^2)] \div 7$
$= [6^2 + (20 \div 5 + 16)] \div 7$
$= [6^2 + (4 + 16)] \div 7$
$= (6^2 + 20) \div 7$
$= (36 + 20) \div 7$
$= 56 \div 7$
$= 8$

20) -14: $(15 \div 5)^2 - [(12 + 2) + 3^2]$
$= (3)^2 - (14 + 3^2)$
$= 9 - (14 + 9)$
$= 9 - 23$
$= -14$

Operations with Fractions

1) $\dfrac{6}{23} + \dfrac{3}{4} = \dfrac{24}{92} + \dfrac{69}{92} = \dfrac{93}{92} = 1\dfrac{1}{92}$

2) $\dfrac{5}{14} + \dfrac{11}{28} = \dfrac{15}{42} + \dfrac{22}{42} = \dfrac{37}{42}$

3) $\dfrac{1}{4} + \dfrac{18}{42} = \dfrac{21}{84} + \dfrac{36}{84} = \dfrac{57}{84} = \dfrac{19}{28}$

4) $\dfrac{9}{10} + \dfrac{18}{35} = \dfrac{63}{70} + \dfrac{36}{70} = \dfrac{99}{70} = 1\dfrac{29}{70}$

5) $\dfrac{18}{27} + \dfrac{4}{6} = \dfrac{24}{54} + \dfrac{36}{54} = \dfrac{60}{54} = \dfrac{10}{9} = 1\dfrac{1}{9}$

6) $\dfrac{1}{2} + \dfrac{2}{3} + \dfrac{3}{4} = \dfrac{6}{12} + \dfrac{8}{12} + \dfrac{9}{12} = \dfrac{23}{12} = 1\dfrac{11}{12}$

7) $\dfrac{4}{10} + \dfrac{3}{4} + \dfrac{1}{2} = \dfrac{8}{20} + \dfrac{15}{20} + \dfrac{10}{20} = \dfrac{33}{20} = 1\dfrac{13}{20}$

8) $\dfrac{1}{3} + \dfrac{5}{10} + \dfrac{3}{5} = \dfrac{10}{30} + \dfrac{15}{30} + \dfrac{18}{30} = \dfrac{43}{30} = 1\dfrac{13}{30}$

9) $\dfrac{1}{2} + \dfrac{2}{3} + \dfrac{6}{10} = \dfrac{15}{30} + \dfrac{20}{30} + \dfrac{18}{30} = \dfrac{53}{30} = 1\dfrac{23}{30}$

10) $\dfrac{1}{3} + \dfrac{8}{10} + \dfrac{1}{4} = \dfrac{20}{60} + \dfrac{48}{60} + \dfrac{15}{60} = \dfrac{83}{60} = 1\dfrac{23}{60}$

11) $\dfrac{4}{8} + \dfrac{1}{3} + \dfrac{6}{16} = \dfrac{24}{48} + \dfrac{16}{48} + \dfrac{18}{48} = \dfrac{58}{48} = \dfrac{29}{24} = \dfrac{15}{24}$

12) $\dfrac{1}{3} + \dfrac{1}{4} + \dfrac{3}{5} = \dfrac{20}{60} + \dfrac{15}{60} + \dfrac{36}{60} = \dfrac{71}{60} = 1\dfrac{11}{60}$

13) $\dfrac{3}{23} + \dfrac{3}{4} + \dfrac{2}{4} = \dfrac{12}{92} + \dfrac{69}{92} + \dfrac{46}{92} = \dfrac{127}{92} = 1\dfrac{35}{92}$

14) $\dfrac{10}{18} + \dfrac{2}{4} + \dfrac{7}{12} = \dfrac{20}{36} + \dfrac{18}{36} + \dfrac{21}{36} = \dfrac{59}{36} = 1\dfrac{23}{36}$

15) $\dfrac{1}{2} + \dfrac{4}{5} + \dfrac{2}{4} = \dfrac{10}{20} + \dfrac{16}{20} + \dfrac{10}{20} = \dfrac{36}{20} = \dfrac{9}{5} = 1\dfrac{4}{5}$

16) $\dfrac{14}{18} - \dfrac{3}{6} = \dfrac{14}{18} - \dfrac{9}{18} = \dfrac{5}{18}$

17) $\dfrac{4}{5} - \dfrac{8}{12} = \dfrac{48}{60} - \dfrac{40}{60} = \dfrac{8}{60} = \dfrac{2}{15}$

18) $\dfrac{1}{4} - \dfrac{2}{14} = \dfrac{7}{28} - \dfrac{4}{28} = \dfrac{3}{28}$

19) $\dfrac{11}{13} - \dfrac{15}{26} = \dfrac{22}{26} - \dfrac{15}{26} = \dfrac{7}{26}$

20) $\dfrac{3}{5} - \dfrac{8}{15} = \dfrac{9}{15} - \dfrac{8}{15} = \dfrac{1}{15}$

21) $\dfrac{16}{24} - \dfrac{1}{6} = \dfrac{16}{24} - \dfrac{4}{24} = \dfrac{12}{24} = \dfrac{1}{2}$

22) $\dfrac{10}{12} - \dfrac{6}{9} = \dfrac{30}{36} - \dfrac{24}{36} = \dfrac{6}{36} = \dfrac{1}{6}$

23) $\dfrac{2}{5} - \dfrac{3}{40} = \dfrac{16}{40} - \dfrac{3}{40} = \dfrac{13}{40}$

24) $\dfrac{6}{55} - \dfrac{1}{11} = \dfrac{6}{55} - \dfrac{5}{55} = \dfrac{1}{55}$

25) $\dfrac{3}{4} - \dfrac{4}{15} = \dfrac{45}{60} - \dfrac{16}{60} = \dfrac{29}{60}$

26) $\dfrac{4}{5} - \dfrac{1}{4} - \dfrac{1}{10} = \dfrac{16}{20} - \dfrac{5}{20} - \dfrac{2}{20} = \dfrac{9}{20}$

27) $\dfrac{8}{10} - \dfrac{1}{4} - \dfrac{1}{5} = \dfrac{16}{20} - \dfrac{5}{20} - \dfrac{4}{20} = \dfrac{7}{20}$

28) $\dfrac{4}{5} - \dfrac{1}{3} - \dfrac{6}{10} = \dfrac{24}{30} - \dfrac{10}{30} - \dfrac{3}{30} = \dfrac{11}{30}$

29) $\dfrac{9}{10} - \dfrac{1}{3} - \dfrac{1}{10} = \dfrac{27}{30} - \dfrac{10}{30} - \dfrac{6}{30} = \dfrac{11}{30}$

30) $\dfrac{4}{5} - \dfrac{1}{3} - \dfrac{1}{5} = \dfrac{8}{10} - \dfrac{5}{10} - \dfrac{2}{10} = \dfrac{1}{10}$

Operations with Fractions

1) $\frac{5}{8} \times \frac{3}{5} = \frac{15}{40} = \frac{3}{8}$

2) $\frac{7}{10} \times \frac{3}{8} = \frac{21}{80}$

3) $\frac{2}{3} \times \frac{3}{8} = \frac{6}{24} = \frac{1}{4}$

4) $\frac{2}{15} \times \frac{6}{7} = \frac{12}{105} = \frac{4}{35}$

5) $\frac{3}{11} \times \frac{5}{12} = \frac{5}{44}$

6) $\frac{2}{3} \times \frac{1}{3} = \frac{2}{9}$

7) $\frac{3}{5} \times \frac{1}{2} = \frac{3}{10}$

8) $\frac{4}{5} \times \frac{4}{7} = \frac{16}{35}$

9) $\frac{1}{2} \times \frac{9}{10} = \frac{9}{20}$

10) $\frac{7}{9} \times \frac{1}{2} = \frac{7}{18}$

11) $\frac{9}{10} \times \frac{3}{14} = \frac{27}{140}$

12) $\frac{8}{15} \times \frac{1}{5} = \frac{8}{75}$

13) $\frac{6}{7} \times \frac{8}{81} = \frac{16}{189}$

14) $\frac{1}{20} \times \frac{7}{25} = \frac{7}{500}$

15) $\frac{1}{5} \times \frac{5}{6} = \frac{5}{30}$

16) $\frac{10}{12} \times \frac{6}{7} = \frac{60}{84} = \frac{5}{7}$

17) $\frac{8}{10} \times \frac{1}{6} = \frac{8}{60} = \frac{2}{15}$

18) $\frac{3}{12} \times \frac{5}{14} = \frac{15}{168} = \frac{5}{56}$

19) $\frac{1}{2} \times \frac{2}{15} = \frac{2}{30} = \frac{1}{15}$

20) $\frac{3}{4} \times \frac{17}{20} = \frac{51}{80}$

21) $\frac{8}{20} \times \frac{5}{8} = \frac{40}{160} = \frac{1}{4}$

22) $\frac{4}{5} \times \frac{9}{12} = \frac{36}{60} = \frac{3}{5}$

23) $\frac{17}{18} \times \frac{4}{10} = \frac{68}{180} = \frac{17}{45}$

24) $\frac{2}{7} \times \frac{9}{15} = \frac{18}{105} = \frac{6}{35}$

25) $\frac{4}{8} \times \frac{13}{18} = \frac{52}{144} = \frac{13}{36}$

26) $\frac{4}{6} \div \frac{6}{9} = \frac{36}{36} = 1$

27) $\frac{2}{5} \div \frac{1}{6} = \frac{12}{5} = 2\frac{2}{5}$

28) $\frac{6}{14} \div \frac{2}{10} = \frac{60}{28} = \frac{15}{7} = 2\frac{1}{7}$

29) $\frac{2}{7} \div \frac{6}{10} = \frac{20}{42} = \frac{10}{21}$

30) $\frac{2}{9} \div \frac{9}{10} = \frac{20}{81}$

Operations with Fractions

1. $\dfrac{1}{9} \div \dfrac{4}{9} = \dfrac{1}{4}$

2. $\dfrac{2}{3} \div \dfrac{5}{7} = \dfrac{14}{15}$

3. $\dfrac{10}{11} \div \dfrac{3}{7} = 2\dfrac{4}{33}$

4. $\dfrac{1}{6} \div \dfrac{3}{5} = \dfrac{5}{18}$

5. $\dfrac{1}{2} \div \dfrac{2}{5} = 1\dfrac{1}{4}$

6. $\dfrac{1}{11} \div \dfrac{3}{7} = \dfrac{7}{33}$

7. $\dfrac{1}{6} \div \dfrac{7}{12} = \dfrac{2}{7}$

8. $\dfrac{2}{9} \div \dfrac{4}{7} = \dfrac{7}{18}$

9. $\dfrac{7}{11} \div \dfrac{1}{4} = 2\dfrac{6}{11}$

10. $5 \div \dfrac{7}{10} = \dfrac{5}{1} \times \dfrac{10}{7} = \dfrac{50}{7} = 1\dfrac{1}{7}$

11. $\dfrac{1}{4} \div 9 = \dfrac{1}{4} \times \dfrac{1}{9} = \dfrac{1}{36}$

12. $10 \div \dfrac{2}{3} = \dfrac{10}{1} \times \dfrac{3}{2} = \dfrac{30}{2} = 15$

13. $\dfrac{1}{5} \div 9 = \dfrac{1}{5} \times \dfrac{1}{9} = \dfrac{1}{45}$

14. $\dfrac{8}{10} \div 3 = \dfrac{8}{10} \times \dfrac{1}{3} = \dfrac{8}{30} = \dfrac{4}{15}$

15. $3 \div \dfrac{1}{5} = \dfrac{3}{1} \times \dfrac{5}{1} = 15$

16. $3\dfrac{5}{9} \div 3\dfrac{3}{7} = \dfrac{32}{9} \times \dfrac{7}{24} = \dfrac{224}{216} = \dfrac{28}{27} = 1\dfrac{1}{27}$

17. $2\dfrac{1}{3} \div 2\dfrac{1}{4} = \dfrac{7}{3} \times \dfrac{9}{4} = \dfrac{28}{27} = 1\dfrac{1}{27}$

18. $3\dfrac{5}{6} \div 4\dfrac{2}{5} = \dfrac{23}{6} \times \dfrac{5}{22} = \dfrac{115}{132}$

19. $2\dfrac{1}{9} \div 3\dfrac{1}{7} = \dfrac{19}{9} \times \dfrac{7}{22} = \dfrac{133}{198}$

20. $2\dfrac{1}{8} \div 2\dfrac{3}{5} = \dfrac{25}{8} \times \dfrac{5}{13} = \dfrac{125}{104} = 1\dfrac{21}{104}$

21. $3\dfrac{1}{3} \div 3\dfrac{3}{4} = \dfrac{10}{3} \times \dfrac{4}{15} = \dfrac{40}{45} = \dfrac{8}{9}$

22. $2\dfrac{1}{2} \div 2\dfrac{3}{7} = \dfrac{5}{2} \times \dfrac{7}{17} = \dfrac{35}{34} = 1\dfrac{1}{34}$

23. $2\dfrac{7}{9} \div 3\dfrac{3}{7} = \dfrac{25}{9} \times \dfrac{7}{24} = \dfrac{175}{216}$

24. $3\dfrac{1}{2} \div 4\dfrac{1}{3} = \dfrac{7}{2} \times \dfrac{3}{13} = \dfrac{21}{26}$

25. $2\dfrac{5}{8} \div 2\dfrac{5}{7} = \dfrac{21}{8} \times \dfrac{7}{19} = \dfrac{147}{152}$

26. $2\dfrac{1}{2} \div 2\dfrac{3}{7} = \dfrac{5}{2} \times \dfrac{7}{17} = \dfrac{35}{34} = 1\dfrac{1}{34}$

27. $3\dfrac{5}{8} \div 3\dfrac{1}{6} = \dfrac{29}{8} \times \dfrac{6}{19} = \dfrac{174}{152} = \dfrac{87}{76} = 1\dfrac{11}{76}$

28. $4\dfrac{2}{5} \div 2\dfrac{6}{7} = \dfrac{22}{5} \times \dfrac{7}{20} = \dfrac{154}{100} = \dfrac{77}{50} = 1\dfrac{27}{50}$

29. $3\dfrac{2}{7} \div 4\dfrac{3}{5} = \dfrac{23}{7} \times \dfrac{5}{23} = \dfrac{115}{161} = \dfrac{5}{7}$

30. $4\dfrac{1}{2} \div 3\dfrac{2}{3} = \dfrac{9}{2} \times \dfrac{3}{11} = \dfrac{27}{22} = 1\dfrac{5}{22}$

Operations with Decimals

1) 62.289 + 33.259 = 95.548

2) 86.927 + 14.969 = 101.896

3) 40.847 + 47.716 = 88.563

4) 95.689 + 67.132 = 162.821

5) 39.323 + 76.324 = 115.647

6) 53.379 + 87.957 = 141.336

7) 65.955 + 86.635 = 152.590

8) 54.557 + 97.499 + 75.548 = 227.604

9) 13.145 + 38.778 + 85.332 = 137.255

10) 80.433 + 13.737 + 88.967 = 183.137

11) 74.197 + 79.562 + 79.967 = 233.726

12) 41.675 + 62.566 + 67.483 = 171.724

13) 36.824 + 88.524 + 66.925 = 192.273

14) 45.152 + 90.349 + 99.487 = 234.988

15) 31.238 + 32.123 + 98.857 = 162.218

16) 991 − 63.609 = 927.391

17) 390 − 34.06 = 355.94

18) 4335 − 9.448 = 4325.552

19) 8592 − 4714.4216 = 3877.5784

20) 8483 − 53.28 = 8429.72

21) 9359 − 284.9 = 9074.1

22) 4706 − 5.4222 = 4700.5778

23) 8352 − 4891.3918 = 3460.6082

24) 8079 − 5.1 = 8073.9

25) 981 − 441.949 = 539.051

26) 755 − 0.1448 = 754.8552

27) 7914 − 2174.557 = 5739.443

28) 2978 − 61.257 = 2916.743

29) 5954 − 3826.56 = 2127.44

30) 660 − 359.2 = 300.8

Operations with Decimals

1) 441 − 57.866 = 383.134

2) 37.79 x 13.6 = 513.9440

3) 66.66 x 28.27 = 1884.4782

4) 12.32 x 68.18 = 839.9776

5) 43.25 x 14.43 = 624.0975

6) 33.79 x 28.59 = 966.0561

7) 75.58 = 25.87 = 1955.2546

8) 44.85 x 30.58 = 1371.513

9) 24.43 x 74.27 = 1814.4161

10) 77.21 x 79.19 = 6114.2599

11) 58.28 x 68.37 = 3984.6036

12) 79.34 x 46.65 = 3701.211

13) 47.87 x 56.24 = 2692.2088

14) 81.12 x 37.83 = 3068.7696

15) 91.26 x 98.37 = 8977.2462

16) 66.86 x 55.76 = 3728.1136

17) 4193.73 ÷ 9 = 465.97

18) 547.04 ÷ 8 = 68.38

19) 309.26 ÷ 94 = 3.29

20) 594.72 ÷ 6 = 99.12

21) 758.08 ÷ 8 = 94.76

22) 5744.96 ÷ 32 = 179.53

23) 1514.72 ÷ 8 = 189.34

24) 95.68 ÷ 13 = 7.36

25) 276.54 ÷ 33 = 8.38

26) 4785.04 ÷ 52 = 92.02

27) 2463.84 ÷ 48 = 51.33

28) 17688.58 ÷ 26 = 680.33

29) 7043.52 ÷ 46 = 153.12

30) 10349.94 ÷ 51 = 202.94

Converting between Fractions and Decimals

1. $\frac{1}{12}$
2. 0.6
3. 0.75
4. 0.667
5. 0.833

6. 0.375
7. 0.091
8. 0.875
9. 0.444
10. 0.944

11. 0.769
12. 0.571
13. 0.053
14. 0.36
15. 0.286

16. $\frac{3}{10}$
17. $\frac{27}{50}$
18. $\frac{1}{5}$
19. $\frac{11}{25}$
20. $\frac{1}{100}$

21. $\frac{49}{50}$
22. $\frac{16}{25}$
23. $\frac{37}{100}$
24. $\frac{1}{8}$
25. $\frac{1}{1}$

Percents

1	2.40	2	45
3	0.52	4	67.53%
5	16.05	6	59.04%
7	88.31	8	103.53
9	223.53	10	16.05
11	19.80	12	4.08
13	74.07	14	3100
15	10.14	16	75.27
17	45.57%	18	12.07
19	16.33%	20	120.51

Ratios

1) 12:15 = 40:50 = 28:35 = 4:5 = 20:25 = 32:40

2) 21:27 = 49:63 = 70:90 = 35:45 = 56:72 = 7:9

3) 40:48 = 25:30 = 35:42 = 15:18 = 5:6 = 50:60

4) 5:9 = 35:63 = 45:81 = 50:90 = 10:18 = 40:72

5) 16:24 = 2:3 = 6:9 = 20:30 = 10:15 = 14:21

6) 30:40 = 21:28 = 9:12 = 35:50 = 6:8 = 24:32

7) 21:30 = 56:80 = 14:20 = 35:50 = 49:70 = 70:100

8) 6:42 = 10:70 = 7:49 = 3:21 = 4:28 = 8:56

9) 35:49 = 15:21 = 50:70 = 10:14 = 25:35 = 40:56

10) 7:28 = 3:12 = 20:80 = 5:20 = 8:32 = 9:36

11) 24:56 = 9:21 = 27:63 = 60:140 = 21:49 = 3:7

12) 4:36 = 8:72 = 3:27 = 9:81 = 2:18 = 4:45

13) 27:45 = 24:40 = 21:35 = 60:100 = 3:5 = 12:20

14) 4:14 = 10:35 = 2:7 = 8:28 = 16:56 = 6:21

15) 30:48 = 45:72 = 5:8 = 15:24 = 10:16 = 35:56

Rates

1. $2.25 per can of soup
2. 25.83 miles per gallon
3. 1.22 pages read per minute
4. $2.29 per pen
5. 1.56 problems per minute
6. 16.67 miles per gallon
7. 1.6 inches per hour
8. $1.54 per cupcake
9. $2.30 per can
10. 1.31 pushups per second
11. $36.25 per chair
12. $2.14 per slice
13. $2.70 per pack of gum
14. $3.17 per headband
15. $35.00 per textbook

Proportions

1) 490/2 = 245 mph   245 x 16 = 3920 miles

2) $2.80/5 = $0.56/can at the bodega
$5.70/9 = $0.63/can at the supermarket so the bodega has a better value

3) 32/8 = 4 riders per minute x 240 minutes = 960 riders

4) 320/5 = 64 per hour   64 x 24 = 1536 cupcakes

5) 25 / 14 = 1.79 min/page x 175 pages = 313.25 minutes = 5 hours and 13 minutes

6) 70 x 2 = 140 people/hour   140 x 9.5 = 1330 people

7) $1.23/3 = $0.41 for Sweeties' pops
$2.15/5 = $0.43 for Darla's pops so Sweeties' pops are a better value

8) 290/6 = 48.33 pairs per hour   48.33 x 40 = 1933.33 pairs, but since pairs are made in whole numbers, 1933 pairs

9) 146 / 9 (there are 9 segments of 20 seconds in 3 minutes)
= 16.2 jumping jacks in 20 seconds

10) 282 / 3 = 94 points/level   94 x 50 levels = 4700 points

Exponents

1. $7^2 = 49$

2. $4^3 = 64$

3. $-(5)^3 = -125$

4. $3^{-3} = \dfrac{1}{27}$

5. $(-2)^{-3} = -\dfrac{1}{8}$

6. $-(8)^3 = -512$

7. $2^{-7} = \dfrac{1}{128}$

8. $7^{-3} = \dfrac{1}{343}$

9. $-(2)^8 = 256$

10. $7^{-3} \times 7^2 = \dfrac{1}{7}$

11. $\dfrac{2^5}{2} = 2^4 = 16$

12. $\dfrac{4^{-2}}{4} = \dfrac{1}{4^3} = \dfrac{1}{64}$

13. $6x^2 \times 3x = 18x^3$

14. $a^2 \times a^5 \times a^3 = a^{10}$

15. $b \times b^{-4} = \dfrac{1}{b^3}$

Scientific Notation

1. 345.1
2. 81,000
3. 92.5
4. 0.007089
5. 300,100
6. 88,820,000
7. 0.00000402834
8. 1,900,010,000
9. 0.6107
10. 0.0287
11. 7.85×10^2
12. 1.6086×10^4
13. 1.908×10^0
14. 4.78×10^{-3}
15. 9.005983×10^6
16. 6.6×10^{-4}
17. 10^1
18. 7.630002×10^6
19. 7×10^{-2}
20. 5.40×10^2

Absolute Value

1) a = {3.33, -3.33}

2) x = {2, 25}

3) y = {4, -4}

4) b = {11.25, -11.25}

5) c = {18, 46}

6) z = {4, -4}

7) k = {26.33, 23.67}

8) v = {-65, 75}

9) z = {18.2, 17.8}

10) a = {-11.83, -12.17}

11) b = {-132, -12}

12) p = {-154, -238}

13) s = {2.33, -6.33}

14) m = {252, -154}

15) x = {24, 48}

Plotting Rational Numbers on the Coordinate Plane

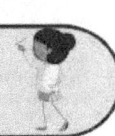

1. (-1, 5)

2. (3, 4)

3. (6, 0)

4. (-4, -1)

5. (-6, 3)

6. (-2, 7)

7. (2, -3)

8. (9, 5)

Plot the Point

1) (2, 4)

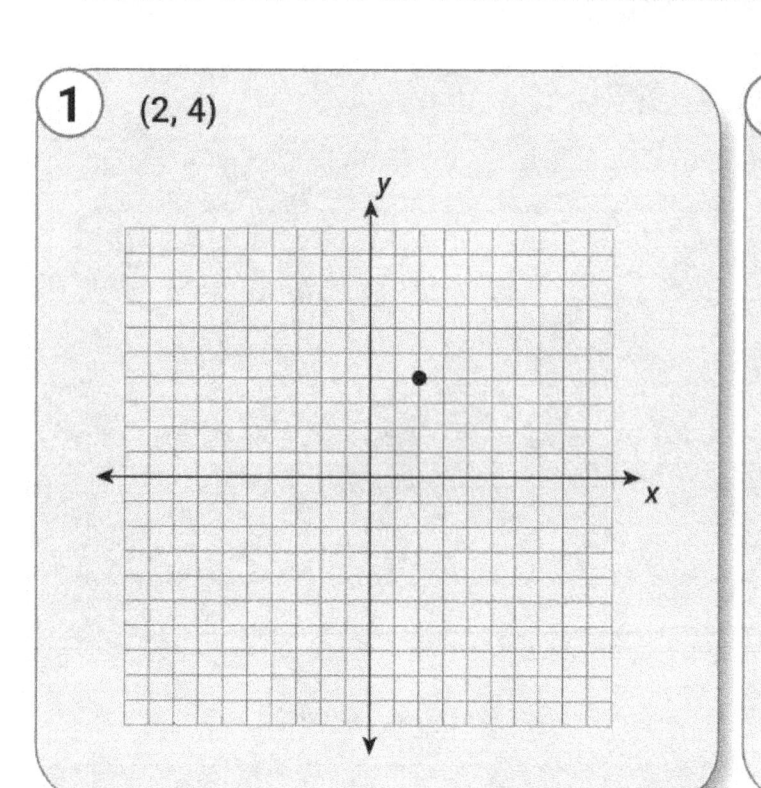

2) (5, 1)

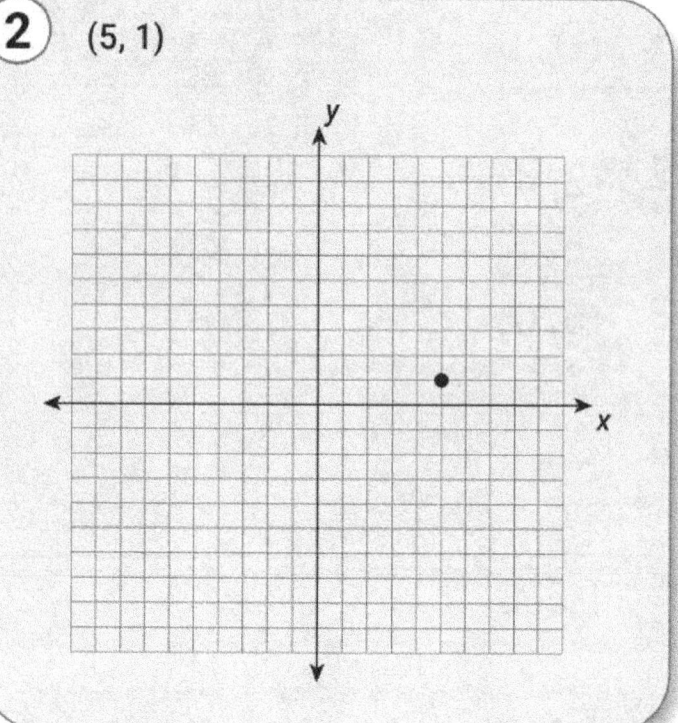

3) (7, 6)

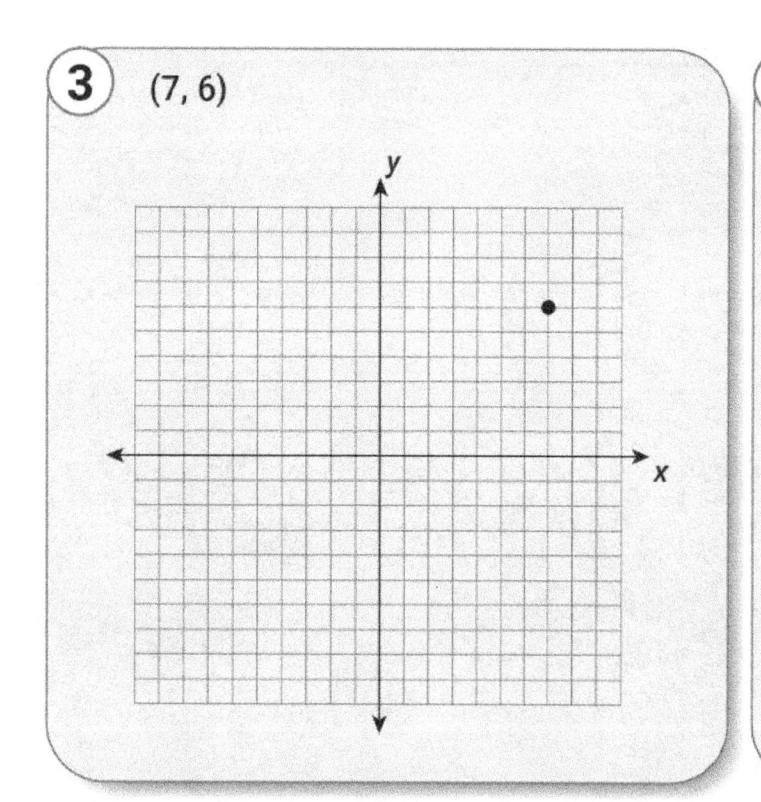

4) (3, -4)

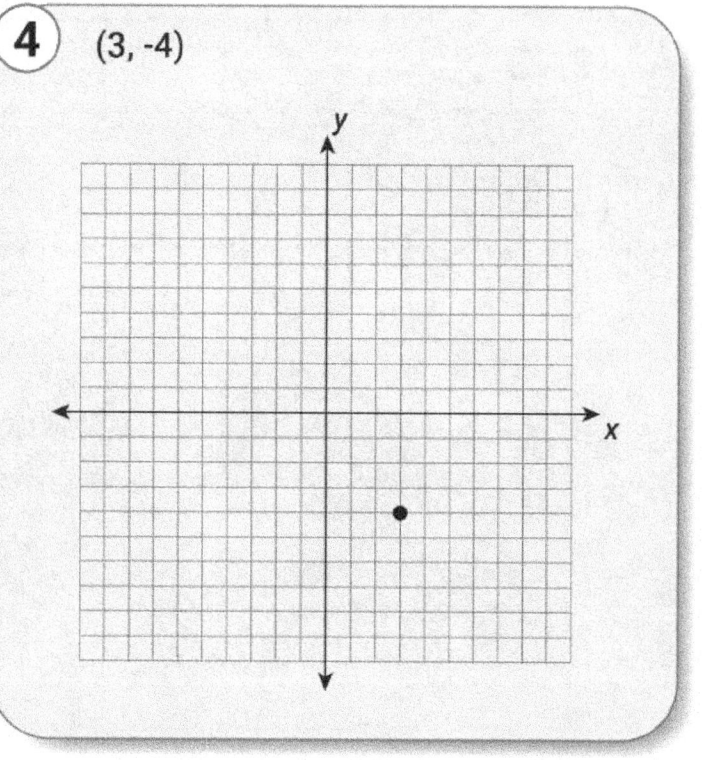

Plot the Point

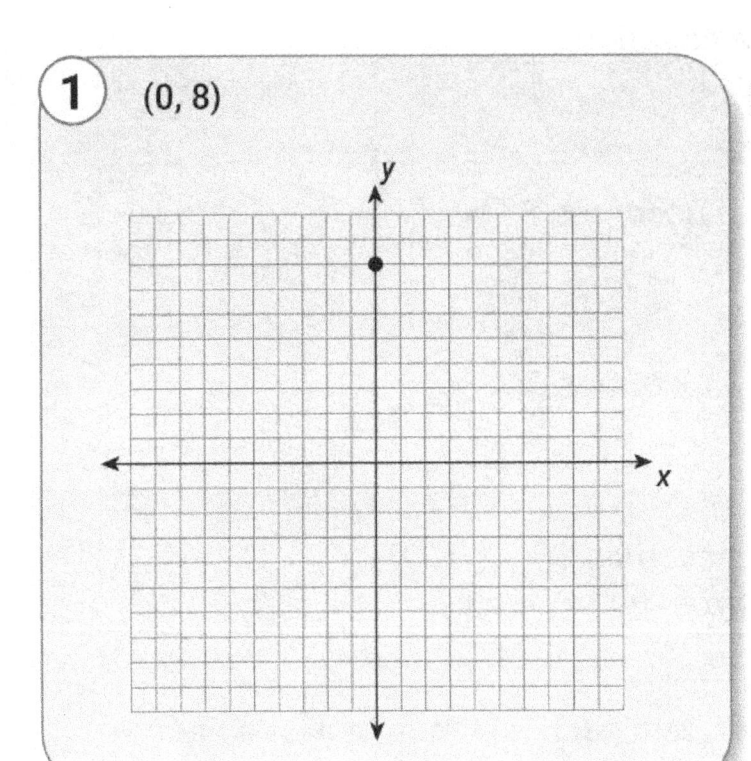

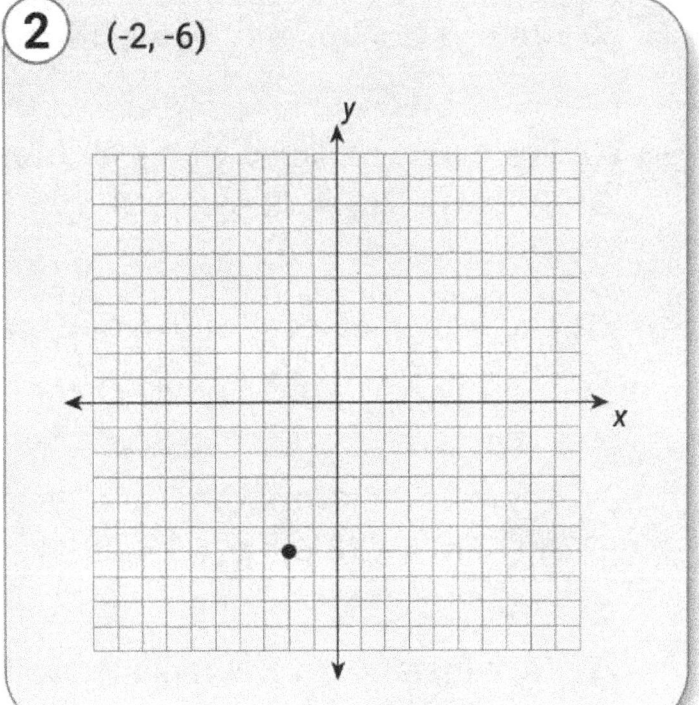

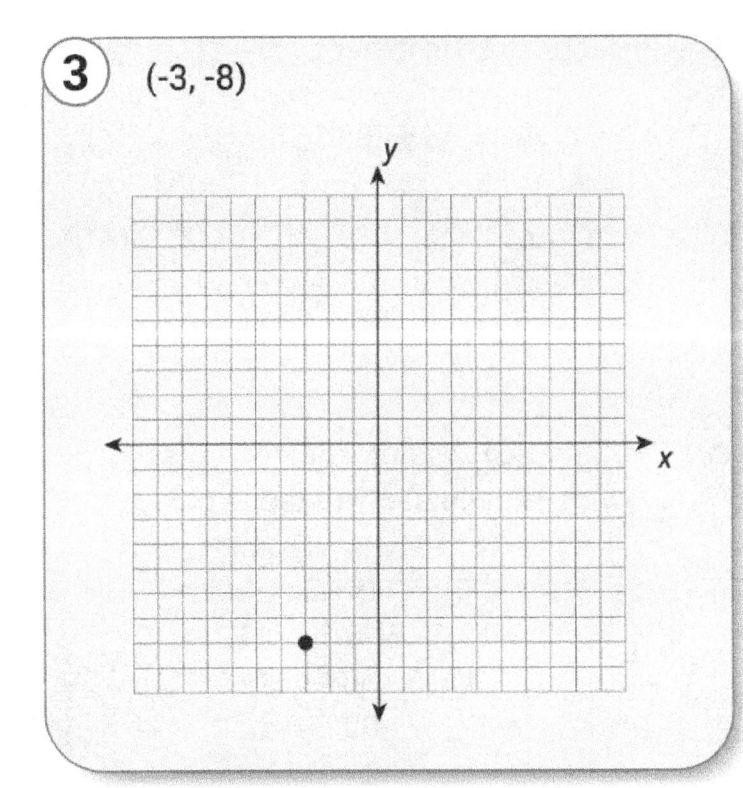

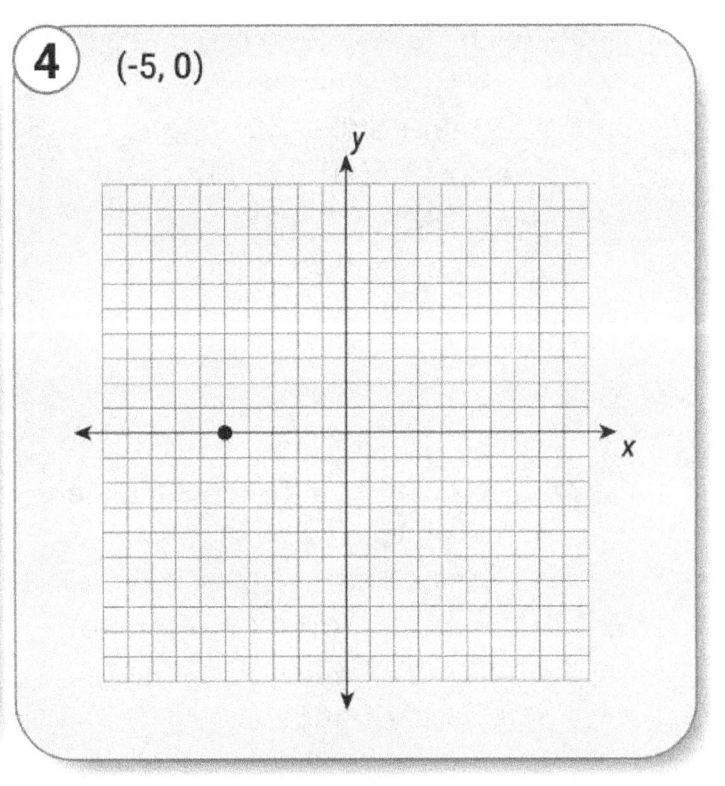

1) (0, 8)

2) (-2, -6)

3) (-3, -8)

4) (-5, 0)

Evaluating Algebraic Expressions

1) 33: Substitute the given values into the equation:
$7b - 2a = 7(7) - 2(8) = 49 - 16 = 33$.

2) 20: The c gets replaced with 4 and the d becomes 8:
$-3 - 9 - 6(4) + 7(8) = -3 - 9 - 24 + 56 = 20$.

3) 51: The x gets replaced with 5 and the y becomes 3:
$6x + 7y = 6(5) + 7(3) = 30 + 21 = 51$.

4) 24: The m is replaced with 3 and the n becomes 6:
$-2(8m - 6n) = -2(8(3) - 6(6)) = -2(24 - 36) = -2(-12) = 24$.

5) 10: Each instance of x is replaced with 2, and each instance of y is replaced with 3 to get $2^2 - 2 \times 2 \times 3 + 2 \times 3^2 = 4 - 12 + 18 = 10$

6) 608: Each instance of n is replaced with 4 to get:
$8n + 5n^3 + 16n^2$
$= 8(4) + 5(4)^3 + 16(4)^2$
$= 32 + 320 + 256$
$= 608$

7) -2028: Each instance of t is replaced with -2, and each instance of g is replaced with 7 to get
$(15 - 8t^2) - (5g^3 - 9 + 6g^2) + (3 + 7t)$
$= (15 - 8(-2)^2) - (5(7)^3 - 9 + 6(7)^2) + (3 + 7(-2))$
$= (15 - 32) - (1715 - 9 + 294) + (3-14)$
$= (-17) - (2000) + (-11)$
$= -2028$

8) -180,072: Each instance of t is replaced with 3, and each instance of b is replaced with -6 to get
$(2a^2 + 6a^4 - 4a)(3b^3 + 8b^2 + b)$
$= (2(3)^2 + 6(3)^4 - 4(3))(3(-6)^3 + + 8(-6)^2 + (-6))$
$= (18 + 486 - 12)(-648 + 288 - 6)$
$= (492)(-366)$
$= -180,072$

9) 1008: Each instance of k is replaced with 12, and each instance of l is replaced with -8 to get
$9k^2 - 6l^2 + 8k$
$= 9(12)^2 - 6(-8)^2 + 8(12)$
$= 1296 - 384 + 96$
$= 1,008$

Evaluating Algebraic Expressions

10) 11: Each instance of b is replaced with 2 to get
$(8b^2 + 3) + (58 - 2b^3) - (6b^3 + 4b)$
$= (8(2)^2 + 3) + (58 - 2(2)^3) - (6(2)^3 + 4(2))$
$= (32 + 3) + (58 - 26) - (48 + 8)$
$= (35) + (32) - (56)$
$= 11$

11) -802: Each instance of c is replaced with -4, and each instance of d is replaced with 3 to get
$(8c^3 - 6c^2 + 4) + (3d^3 + 7c^2)$
$= (8(-4)^3 - 6(-4)^2 + 4) + (3(3)^3 + 7(3)^2)$
$= (-512 - 384 + 4) + (27 + 63)$
$= (-892) + (90)$
$= -802$

12) -3948: Each instance of x is replaced with -3, and each instance of y is replaced with 7 to get
$y(9 - 7x^4 + 2x)$
$= 7(9 - 7(-3)^4 + 2(-3))$
$= 7(9 - 567 - 6)$
$= 7(-564)$
$= -3,948$

13) 828: Each instance of c is replaced with 14, and each instance of d is replaced with 4 to get
$(8 + 4c^2) - (2d^3 - 3d^2)$
$= (8 + 4(15)^2) - (2(4)^3 - 3(4)^2)$
$= (908) - (128 - 48)$
$= 828$

14) 55: Each instance of m is replaced with 1, and each instance of n is replaced with -1 to get
$(2m^2 + 3)(4n^2 - 7n)$
$= (2(1)^2 + 3)(4(-1)^2 - 7(-1))$
$= (5)(11)$
$= 55$

Solving Equations

1) -5x = -40. Divide each side by -5. X = 8

2) 2 + j = -8. Subtract 2 from both sides, which yields j = -10.

3) -7 + a = -10. Add 7 to both sides, which gives a = -3.

4) -8h + 6h = 22. Simplify the left side of the equation first: -2h = 22. Then divide both sides by -2: h = -11.

5) 12 = m − 2. Add 2 to both sides, which gives the value m = 14.

6) 11 = c − 3. Add 3 to both sides, which gives the value of c = 14.

7) -12 = 2y. Divide both sides by 2. Then y = -6.

8) 7f = 56. Divide both sides by 7, which yields f = 8.

9) -34 = 6.8c. Divide both sides by 7, which yields c = -5.

10) 6g = 36. Divide both sides by 6, which yields g = 6.

11) $\frac{z}{4}$ = 7. Multiply both sides by 4 to get rid of the fraction. This leaves z = 28.

12) -5.1d = -35.7. Divide both sides by -5.1: d = 7.

13) $\frac{b}{3}$ = -4. Multiply both sides by 3 to get rid of the fraction. This leaves b = -12.

14) 2.5t = 10. Divide both sides by 2.5, which yields t = 4.

15) 46.4 = -5.8a. Divide both sides by -5.8, which yields a = -8.

16) $\frac{j}{5}$ = 4.2. Multiply both sides by 5: j = 21.

17) 31.8 = 5.3x. Divide both sides by 5.3, which yields x = 6.

18) -16 = -2b -4 + 5b. Simplify the right side of the equation: -16 = 3b -4. Then add 4 to both sides and then divide both sides by -3 to isolate the variable. The result is b = -4.

19) 9 = -23u − 22. Add 22 to both sides and then divide both sides by -23 to isolate the variable. The result is u = -1.35.

20) 23 = 13d + 2. Subtract 2 from both sides and then divide both sides by 13 to isolate the variable. The result is d = 1.62.

Solving Equations

1) $\frac{5}{6}$m - 19 = -5. Add 19 to both sides and then multiply both sides by $\frac{6}{5}$ (the reciprocal of the coefficient of m) to isolate the variable. The result is m = 16.8.

2) $\frac{1}{7}$v + 19 = 21. Subtract 19 from both sides and then multiply both sides by 7 (the reciprocal of the coefficient of v) to isolate the variable. The result is v = -14.

3) 7 = $\frac{f-4}{5}$. Multiply both sides by 5 to get rid of the fraction. Then add 4 to both sides to isolate the variable. The result is f = 39.

4) $\frac{2+12}{3}$ = 11. Multiply both sides by 3 to get rid of the fraction. Then subtract 12 from both sides to isolate the variable. The result is w = 21.

5) -7q - 9q = -32. Subtract 9q from -7q, which gives -16q = -32. Then, divide both sides by -16, which results in q = 2.

6) -10 = $\frac{v-27}{-24}$. Multiply both sides by -24 to get rid of the fraction. Then add 27 to both sides to isolate the variable. The result is p = 267.

7) $\frac{2}{5}$n + 21 = 18. Subtract 21 from both sides. Then multiply both sides by $\frac{5}{2}$, which is the reciprocal of the coefficient of n. Doing so, isolates the variable n with the value of -7.5.

8) $\frac{p-14}{8}$ = 28. Multiply both sides by 8 to get rid of the fraction. Then add 14 to both sides to isolate the variable. The result is p = 238.

9) -25 = $\frac{28-k}{3}$. Multiply both sides by 3 to get rid of the fraction. Then subtract k from both sides to isolate the variable. The result is -k = -103, which means that k = 103.

10) 4(x - 7) - 8 = 30. Add 8 to both sides, which yields 4(x - 7) = 38. Then, divide both sides by 4, which results in x - 7 = 9.5. Lastly, add 7 to both sides to isolate the variable: x = 16.5.

11) $\frac{7+r}{26}$ = 23. Multiply both sides by 26 to get rid of the fraction. Then subtract 7 from both sides to isolate the variable. The result is r = 591.

12) 2(7 - 3d) = 23. Divide both sides by 2, which yields 7 - 3d = 11.5. Then, subtract 7 from both sides and divide both sides by -3 to isolate the variable. The result is d = -1.5.

13) Add 3 to both sides to get 4x = 8. Then divide both sides by 4 to get x = 2.

14) First, subtract 4 from each side. This yields 6x = 12. Now, divide both sides by 6 to obtain x = 2.

15) Start by squaring both sides to get 1 + x = 16. Then subtract 1 from both sides to get x = 15.

16) Multiply both sides by x to get x + 2 = 2x, which simplifies to -x = -2, or x = 2.

17) The first step in solving this equation is to collect like terms on the left side of the equation. This yields the new equation -4 + 8x = 8 - 10x. The next step is to move the x-terms to one side by adding 10 to both sides, making the equation -4 + 18x = 8. Then the -4 can be moved to the right side of the equation to form 18x = 12. Dividing both sides of the equation by 18 gives a value of 0.67, or $\frac{2}{3}$.

18) x = 150. Set up the initial equation:
$\frac{2x}{5}$ - 1 = 59
Add 1 to both sides:
$\frac{2x}{5}$ - 1 + 1 = 59 + 1
Multiply both sides by $\frac{5}{2}$:
$\frac{2x}{5}$ × $\frac{5}{2}$ = 60 × $\frac{5}{2}$ = 150
x = 150

19) 25 = 5(3 + x)
25 = 15 + 5x
10 = 5x; x = 2

20) 12 = 2(3x + 4)
12 = 6x + 8
4 = 6x; x = $\frac{2}{3}$

Adding and Subtracting Polynomials

1) $(9 - 7x^4)-(2x^4 + 3) = -9x^4 + 6$

2) $(8 + 4d^2)-(2d^3 - 6 - 3d^2) = -2d^3 + 7d^2 + 14$

3) $(2m^2 + 3)-(4m^2 - 7 + m^4) = -m^4 - 2m^2 + 10$

4) $(x^3 - 3x^2 + 2x - 2)-(3x^3 + 4x - 3) = -2x^3 - 3x^2 - 2x + 1$
The implied +1 in front of the first set of parentheses will not change those four terms; however, distributing the implied -1 in front of the second set of parentheses will change the sign of each of those three terms:
$x^3 - 3x^2 + 2x - 2 - 3x^3 - 4x + 3$
Combining like terms yields:
$-2x^3 - 3x^2 - 2x + 1$

5) $5x^2 - 3x + 4$. By distributing the implied one in front of the first set of parentheses and the − 1 in front of the second set of parentheses, the parenthesis can be eliminated: $1(5x^2 - 3x + 4) - 1(2x^2 - 7) = 5x^2 - 3x + 4 - 2x^2 + 7$
Next, like terms (same variables with same exponents) are combined by adding the coefficients and keeping the variables and their powers the same:
$5x^2 - 3x + 4 - 2x^2 + 7 = 3x^2 - 3x + 11$

6) $(7n + 3n^3 + 3)+(8n + 5n^3 + 2n^4) = 2n^4 + 8n^3 + 15n + 3$

7) $(4 - 8t^2)-(5t^3 - 9 + 6t^2)+(3 + 7t) = -5t^3 - 14t^2 + 7t + 16$

8) $(2 - 9c^3 + 5c^4)-(8c^4 - 6c^2 + 4)-(3c^3 + 7c^2) = -3c^4 - 12c^3 - c^2 - 2$

9) $(5a^2 - 7a + 2)-(8a - 9)+(4a^4 + 6a^2 + 3) = 15a^2 - 15a + 14$

10) $(2a^2 + 6a^4 - 4a)-(3a^2 + 8a^3 + a)+(7a^3 - 5a^4 - 9a^2) = a^4 - a^3 - 10a^2 - 5a$

11) $(2k - 5k^4)+(7k + 4k^3)-(9k^3 - 6k^4 + 8k) = k^4 - 5k^3 + k$

12) $(8b^5 + 3)+(5b^5 + 8 - 2b^3)-(6b^3 + 4b) = 13b^5 - 8b^3 - 4b + 11$

Multiplying Monomials and Polynomials

1 $2(7x^2 + 3x - 8) = 14x^2 + 6x - 16$

2 $4x^2(-x + 3) = -4x^3 + 12x$

3 $6x(7x^2 - 3x + 1) = 42x^3 - 18x^2 + 6x$

4 $9x^3(2x^2 - 7x - 6) = 18x^5 - 63x^4 - 54x^3$

5 $6(3r - 4) = 18r - 24$

6 $12(3x^2 + x) = 36x^2 + 12x$

7 $y^2(y + 15) = y^3 + 15y^2$

8 $5x(x + 4) = 5x^2 + 20x$

9 $(x + 2)(x - 3)(x - 3) = x^3 - 4x^2 - 3x + 18$

Multiplying Monomials and Polynomials

1) $2x^2(7x^2 - 9x + 2) = 14x^4 - 18x^3 + 4x^2$

2) $3x(x^2 + xy + 4y^2) = 3x^3 + 3x^2y + 12xy^2$

3) $6x^3(3x^2 + 2xy + 5y^2) = 18x^5 + 12x^4y + 30x^2y^2$

4) $(4x + 2)^2 = 16x^2 + 16x + 4$

5) $(x + 4)^2 = x^2 + 8x + 16$

6) $(x - 5)(x + 5) = x^2 - 25$

7) $(6x - 9)(6x + 9) = 36x^2 - 81$

8) $(8x + 3)^2 = 64x^2 + 48x + 9$

9) $(x + 2)(x^2 + 5x - 6)$
   $= x^3 + 7x^2 + 4x - 12$
   $= x^3 + 5x^2 - 6x + 2x^2 + 10x - 12$
   $= x^3 + 7x^2 + 4x - 12$

The FOIL Method

1. $(x - 4)(x + 9) = x^2 + 9x + -4x + -36 = x^2 + 5x - 36$

2. $(x - 7)(x - 5) = x^2 + -5x + -7x + 35 = x^2 - 12x + 35$

3. $(x + 8)(x + 1) = x^2 + x + 8x + 8 = x^2 + 9x + 8$

4. $(x + 6)(x + 3) = x + 3x + 6x + 18 = x^2 + 9x + 18$

5. $(x - 10)(x - 2) = x^2 + -2x + -10x + 20 = x^2 - 12x + 20$

6. $(x + 10)(x + 4) = x^2 + 14x + 40$

7. $(x - 2)(x + 8) = x^2 + 8x + -2x + -16 = x^2 + 6x - 16$

8. $(x - 3)(x + 12) = x^2 + 12x + -3x + -36 = x^2 + 9x - 36$

9. $(x + 15)(x - 1) = x^2 + -x + 15x + -15 = x^2 + 14x - 15$

10. $(x + 10)(x + 6) = x^2 + 6x + 10x + 60 = x^2 + 16x + 60$

Factoring

1. $(5a^2 + 5a) = 5a(a + 1)$

2. $t^2 - 3t = t(t - 3)$

3. $a^2 + 7a + 12 = (a + 4)(a + 3)$

4. $3w^2 - 12w - 135 = 3(w + 5)(w - 9)$

5. $(a^2 - 9) = (a - 3)(a + 3)$

6. $(b^2 - 5b) = b(b - 5)$

7. $(x^2 - 16) = (x - 4)(x + 4)$

8. $(t^2 + 4t) = t(t + 4)$

9. $k^2 - k - 56 = (k + 7)(k - 8)$

10. $16y^2 - 60y + 56 = 4(y - 2)(4y - 7)$

Factoring

1. $6r^2 + 31r + 40 = (3r + 8)(2r + 5)$

2. $30v^2 - 105v - 135 = 15(v + 1)(2v - 9)$

3. $6d^2 + 2d - 28 = 2(d - 2)(3d + 7)$

4. $24f^2 + 12f - 36 = 12(f - 1)(2f + 3)$

5. $t^2 + 11t + 18 = (t + 9)(t + 2)$

6. $5p^2 + 5p - 150 = 5(p - 5)(p + 6)$

7. $40c^2 - 36c - 36 = 4(5c + 3)(2c - 3)$

8. $(100y^2 - 36) = (10y - 6)(10y + 6)$

9. $3g^2 + 12g - 36 = 3(g - 2)(g + 6)$

10. $(25x^2 - 64) = (5x - 8)(5x + 8)$

Finding Zeros of Polynomials

1) $x = \frac{2}{3}, 3$: The given equation, $y = (3x - 2)(x - 3)$, is already in factored form, so to find the zeros, each factor just needs to be set equal to zero and solved.

Therefore,
$(3x - 2) = 0$
$3x = 2;\ x = \frac{2}{3}$
And:
$(x - 3) = 0$
$x = 3$

Thus, the equation has zeros, or x-intercepts, at $\frac{2}{3}$ and 3.

2) $x = \frac{5}{3}, -5$: The given equation, $y = (3x - 5)(x + 5)$, is already in factored form, so to find the zeros, each factor just needs to be set equal to zero and solved.

Therefore,
$(3x - 5) = 0$
$3x = 5;\ x = \frac{5}{3}$
And:
$(x + 5) = 0$
$x = -5$

Thus, the equation has zeros, or x-intercepts, at $\frac{5}{3}$ and -5.

3) $x = \frac{4}{5}$ and 7: To find the zeros, the equation, $y = 5x^2 - 39x + 28$, first needs to be factored:

$y = (5x - 4)(x - 7)$

Then, each factor is set equal to zero to solve for x:
$0 = (5x - 4)$
$4 = 5x;\ x = \frac{4}{5}$
And:
$(x - 7) = 0;\ x = 7$

So the x-intercepts are at $\frac{4}{5}$ and 7.

4) $x = 0, -2$: This particular equation can be factored into:

$y = x(x^2 + 4x + 4)$
$x(x + 2)(x + 2)$

By setting each factor equal to zero and solving for x, there are two solutions, $x = 0, -2$. On a graph, these zeros can be seen where the line crosses the x-axis.

Finding Zeros of Polynomials

5. $x = 4, -5$: $y = (x^2 + x - 20)$ can be factored into:
$$y = (x - 4)(x + 5)$$
Then, setting each factor equal to zero yields the x-intercepts of 4 and -5.

6. $x = 0, -1, 4$: Again, finding the zeros for a function by factoring is done by setting the equation equal to zero, then completely factoring. Since there was a common x for each term in the provided equation, that is factored out first. Then the quadratic that is left can be factored into two binomials: $(x + 1)(x - 4)$. Setting each factor equation equal to zero and solving for x yields three zeros.
$$0 = x(x + 1)(x - 4); x = 0, -1, 4.$$

7. $\frac{4}{5}$ and $-\frac{3}{4}$: The polynomial $y = (20x^2 - x - 12)$ factors into:
$$y = (4x + 3)(5x - 4)$$
Setting each factor equal to zero and then solving for x gives us our x-intercepts:
$$0 = (4x + 3)$$
$$-3 = 4x; x = -\frac{3}{4}$$
And:
$$0 = (5x - 4)$$
$$4 = 5x; x = \frac{4}{5}$$

8. $x = \frac{2}{5}, -\frac{3}{5}$: The polynomial $y = (10x^2 - 19x - 15)$ factors into:
$$y = (5x + 3)(2x - 5)$$
Setting each factor equal to zero and then solving for x gives us our x-intercepts:
$$0 = (5x + 3)$$
$$-3 = 5x; x = -\frac{3}{5}$$
And:
$$0 = (2x - 5)$$
$$5 = 2x; x = \frac{2}{5}$$

Finding Zeros of Polynomials

9) x = 2, -4, 3: We need to factor the polynomial y = (x^3 + $5x^2$ - 2x - 24) into (x - 2)(x + 4)(x + 3). From there, we can set each factor equal to zero and solve for x, yielding three roots: x = 2, -4, 3.

10) x = 2, -3, 2: It takes a few steps to factor the polynomial y = (x^2 + 2x - 8).
First, we will group it into:
y = (x^3 + $2x^2$) + (-9x - 18)
Then, we will factor each group:
y = x^2 (x + 2) +-9(x + 2)
= (x^2 - 9)(x - 2)
Then, we can use the difference of squares formula to factor the first factor further, yielding:
y = (x - 3)(x + 3)(x + 2)
Setting each factor equal to zero and then solving for x gives us our x-intercepts: x = 2, -3, 2.

11) x = 3, 4: The polynomial y = (x^2 - 7x + 12) factors into:
y = (x - 3)(x - 4)
From there, we can set each factor equal to zero and solve for x, which allows us to determine the roots to be x = 3, 4.

12) x = $\frac{2}{5}$, $\frac{4}{5}$: The polynomial y = ($25x^2$ - 30x + 8) factors into:
y = (5x - 2)(5x - 4)
Setting each factor equal to zero and solving for x yields:
0 = (5x - 2)
2 = 5x; x = $\frac{2}{5}$
And:
0 = (5x - 4)
4 = 5x; x = $\frac{4}{5}$
So, the zeros of the equation are x= $\frac{2}{5}$, $\frac{4}{5}$.

Finding Zeros of Polynomials

13) $x = \frac{3}{4}, \frac{4}{3}$: The polynomial $y = (12x^2 - 25x + 12)$ factors into $y = (4x - 3)(3x - 4)$.
Then, we set each factor equal to zero and solve for x to find the roots:
$$0 = (4x - 3)$$
$$3 = 4x; \; x = \frac{3}{4}$$
And:
$$0 = (3x - 4)$$
$$4 = 3x; \; x = \frac{4}{3}$$
Therefore, our two roots are $x = \frac{3}{4}, \frac{4}{3}$.

14) $x = -3, -5, -4$: The polynomial $y = (x^3 + 12x^2 + 47x + 60)$ factors into:
$$y = (x + 3)(x + 5)(x + 4)$$
Therefore, after setting each factor equal to zero, we can determine the roots to be $x = -3, -5, -4$.

15) $x = -\frac{4}{5}, \frac{2}{5}$: The polynomial $y = (25x^2 + 10x - 8)$ factors into $y = (5x + 4)(5x - 2)$.
Then, we need to set each factor equal to zero and solve for x:
$$0 = (5x - 2)$$
$$2 = 5x; \; x = \frac{2}{5}$$
And:
$$0 = (5x + 4)$$
$$-4 = 5x; \; x = -\frac{4}{5}$$
Therefore, our roots are $x = -\frac{4}{5}, \frac{2}{5}$.

Determining Slope from Coordinate Pairs

1. slope = 1
2. slope = $\frac{2}{3}$
3. slope = $-\frac{1}{3}$
4. slope = $\frac{1}{10}$
5. slope = 5
6. slope = 1
7. slope = 10
8. slope = 2
9. slope = -6
10. slope = 10

Finding Linear Equations from Graphs

1.

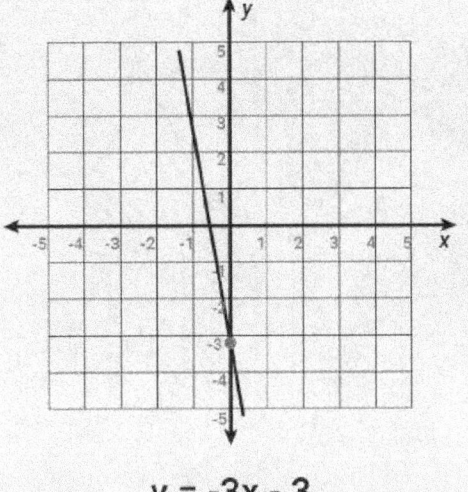

$y = -3x - 3$

2.

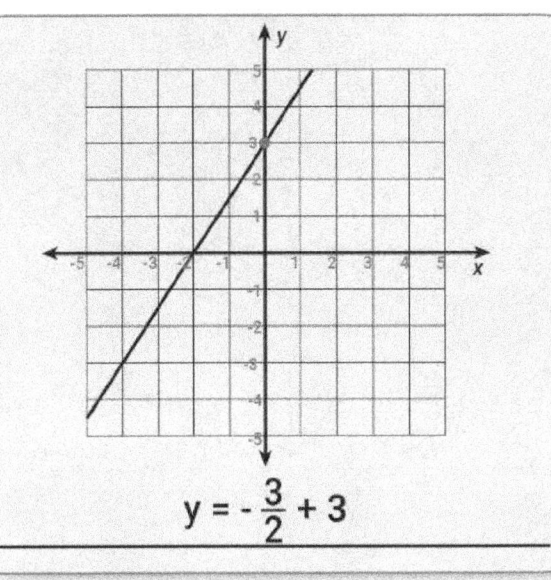

$y = -\frac{3}{2} + 3$

3.

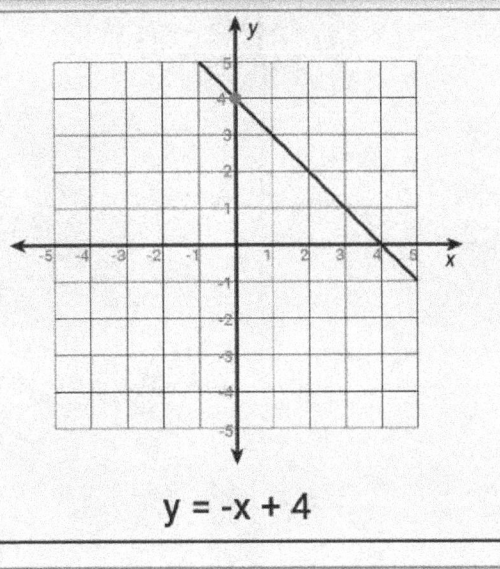

$y = -x + 4$

4.

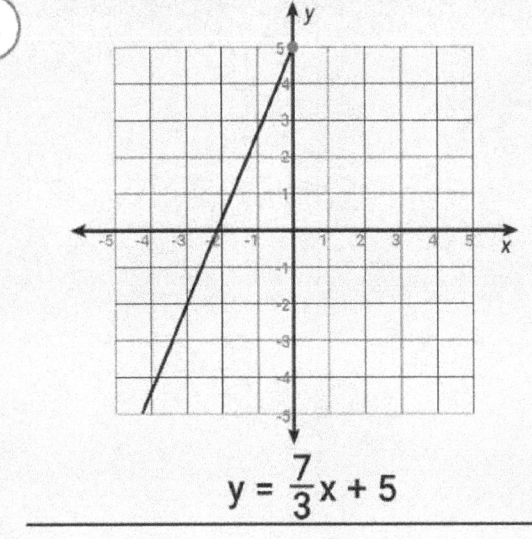

$y = \frac{7}{3}x + 5$

5.

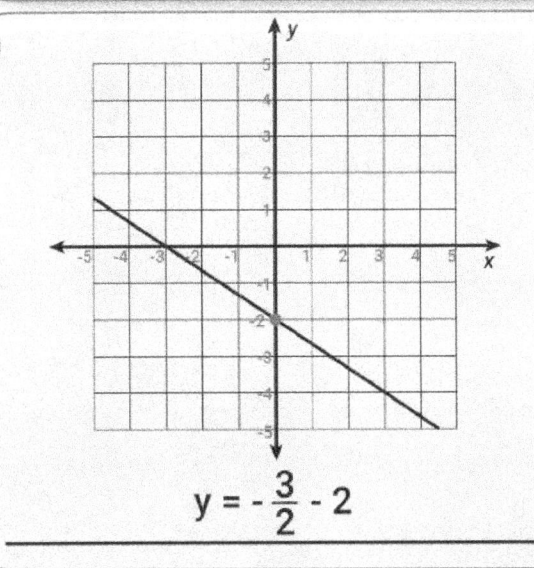

$y = -\frac{3}{2} - 2$

Finding Linear Equations from Graphs

1. 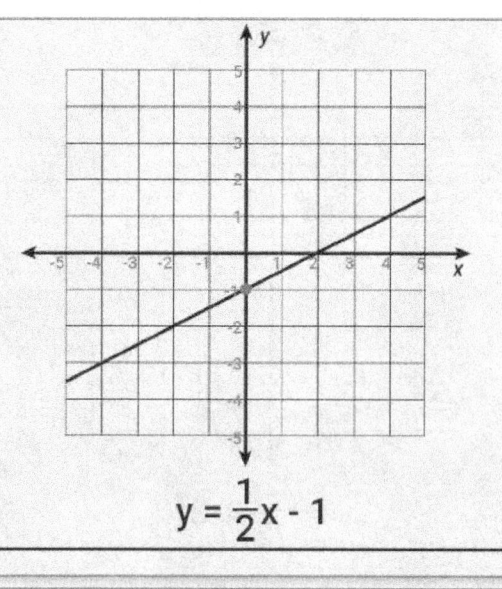 $y = \dfrac{1}{2}x - 1$

2. $y = 2x - 2$

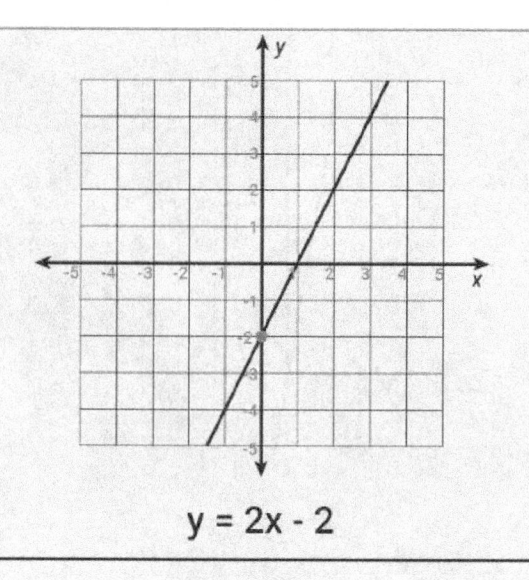

3. $y = \dfrac{1}{2}x - 1$

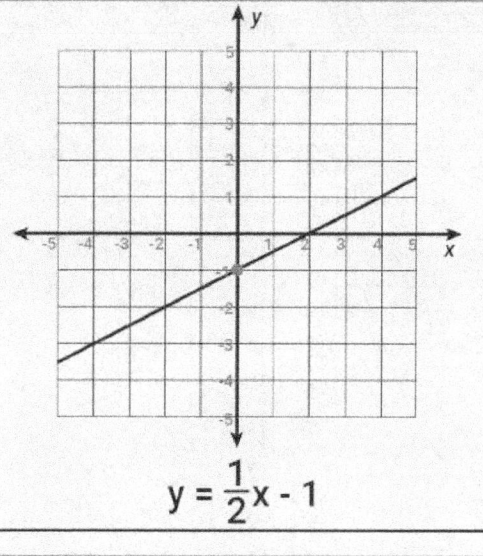

4. $y = -\dfrac{3}{7}x + 4$

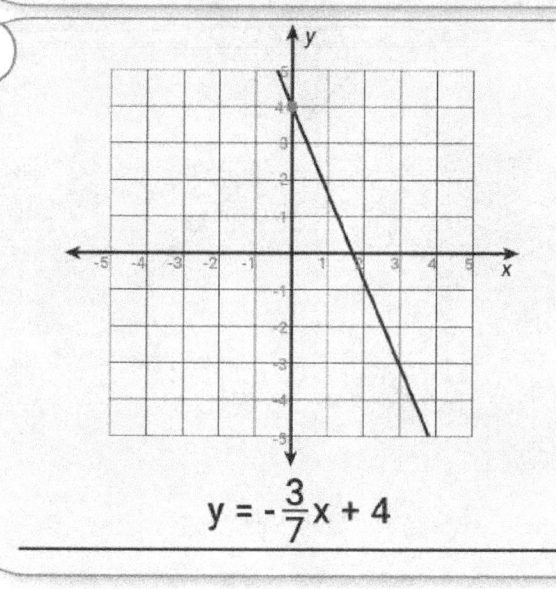

5. $y = -\dfrac{7}{6}x + 2$

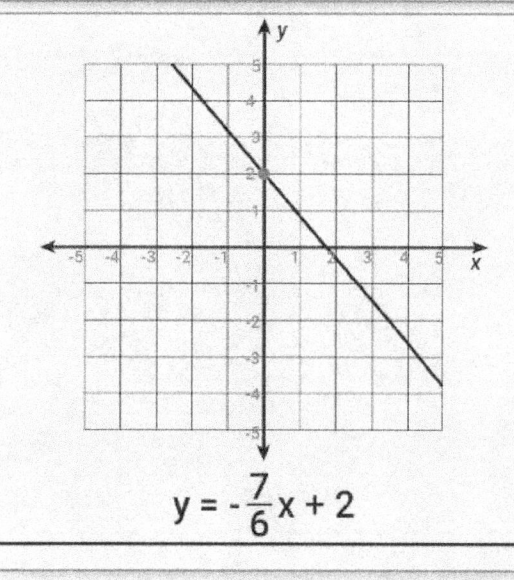

Finding Linear Equations from Graphs

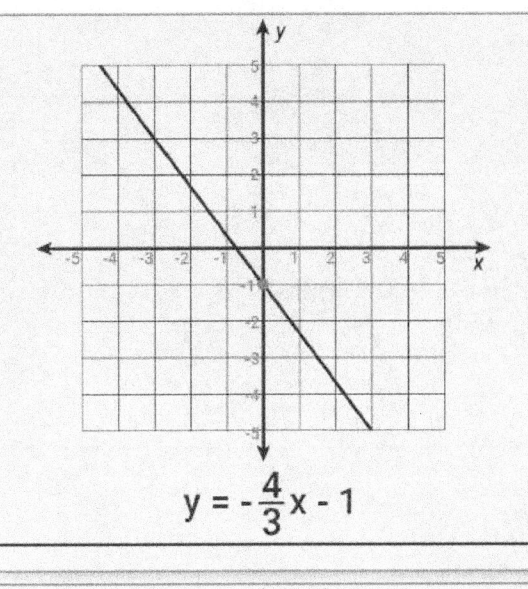

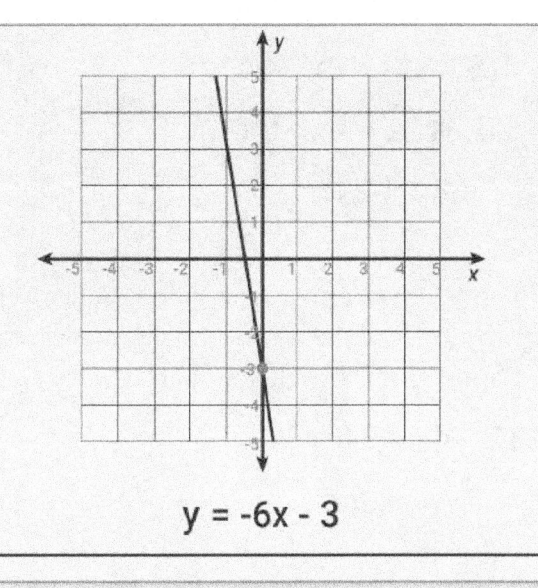

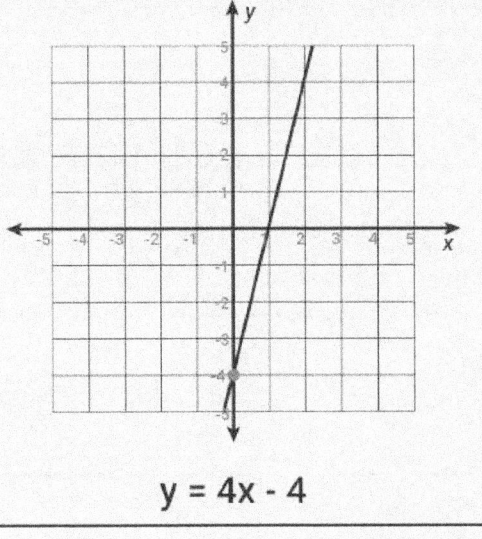

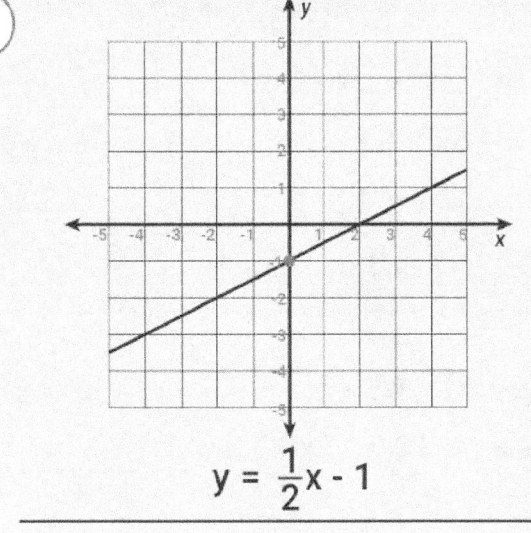

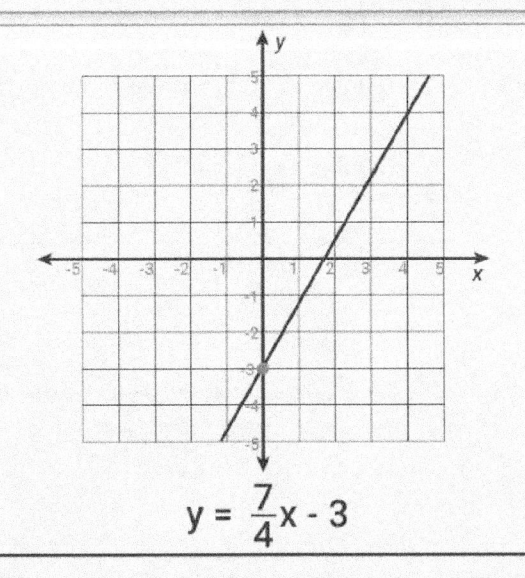

1

$y = -\dfrac{4}{3}x - 1$

2

$y = -6x - 3$

3

$y = 4x - 4$

4

$y = \dfrac{1}{2}x - 1$

5

$y = \dfrac{7}{4}x - 3$

Solving Linear Inequalities

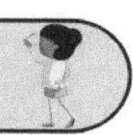

1)
$4(5 - 3x) < 9x - 64$
$20 - 12x < 9x - 64$
$84 < 21x$
$x > 4$

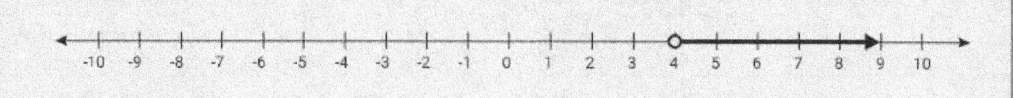

2)
$2(6 - 2x) \leq 16 - 3(x + 4)$
$12 - 4x \leq 16 - 3x - 12$
$x \geq 8$

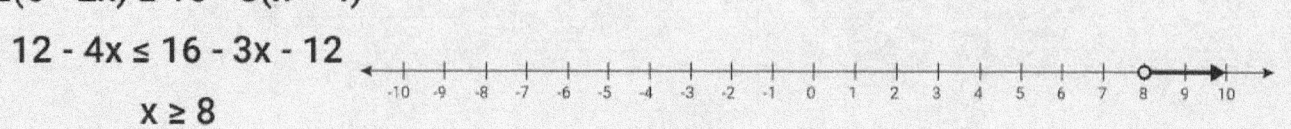

3)
$2(4 - 3x) \geq 8x - 34$
$8 - 6x \geq 8x - 34$
$14x \leq 42$
$x \leq 3$

4)
$-5(1 - x) > 63 - 3(x + 4)$
$-5 + 5x > 63 - 3x - 12$
$8x > 56$
$x > 7$

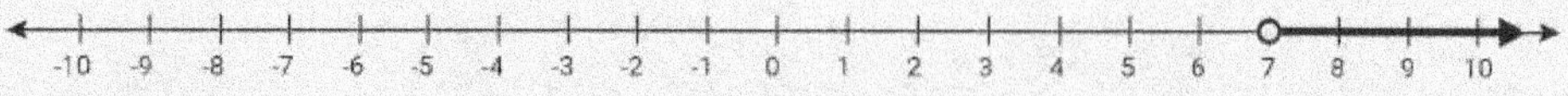

5)
$8(3 + 2x) - x > -3(x + 4)$
$24 + 16x > -3x - 12$
$18x > -36$
$x > -2$

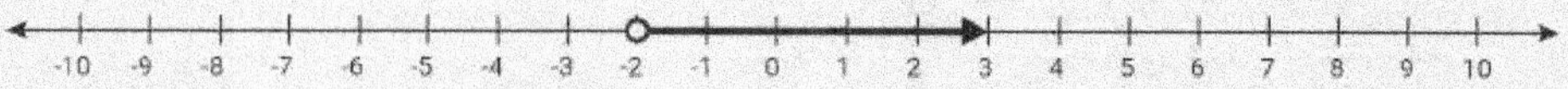

Solving Linear Inequalities

6) $-26 > -3x - 6 + 2x$
$-20 > -x$
$x > 20$

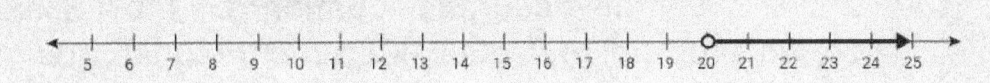

7) $48 > 5x - 8 + 2x$
$56 > 7x$
$x < 8$

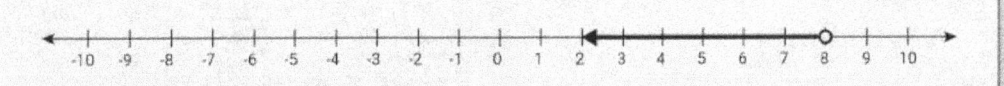

8) $3(6 - 2x) - 4x > 4x - 52$
$18 - 6x - 4x > 4x - 52$
$70 > 14x$
$x < 5$

9) $5x - 146 < 2(3 - 5x) - 4x$
$5x - 146 < 6 - 10x - 4x$
$19x < 152$
$x < 8$

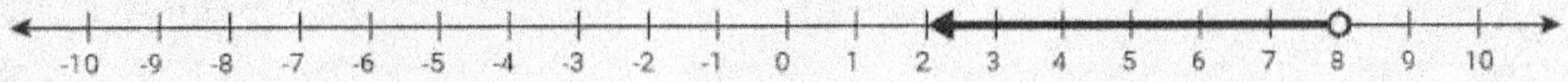

10) $6x - 5 + 3x \geq -77$
$9x \geq -72$
$x \geq -8$

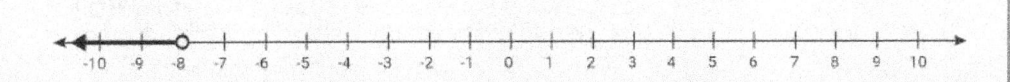

Solving Quadratic Equations Exercises

1

$$x^2 + 3x - 40 = 0$$

For this quadratic equation, a = 1, b = 3, and c = -40.
Then, we simply plug these values into the quadratic formula: $x = \dfrac{-b \pm \sqrt{b^2 - 4ac}}{2a}$

$$x = \dfrac{-3 \pm \sqrt{3^2 - 4(1)(-40)}}{2(1)}$$

$$x = \dfrac{-3 \pm \sqrt{169}}{2}$$

$$x = \dfrac{-2 \pm 13}{2}$$

x = 5, or x = -8

2

$$x^2 - 2x - 120 = 0$$

For this quadratic equation, a = 1, b = -2, and c = -120.
Then, we simply plug these values into the quadratic formula: $x = \dfrac{-b \pm \sqrt{b^2 - 4ac}}{2a}$

$$x = \dfrac{-(-2) \pm \sqrt{(-2)^2 - 4(1)(-120)}}{2(1)}$$

$$x = \dfrac{2 \pm \sqrt{4-(-480)}}{2}$$

$$x = \dfrac{2 \pm 22}{2}$$

x = 12, or x = -10

3

$$x^2 - 20x - 96 = 0$$

For this quadratic equation, a = 1, b = -20, and c = -96.
Then, we simply plug these values into the quadratic formula: $x = \dfrac{-b \pm \sqrt{b^2 - 4ac}}{2a}$

$$x = \dfrac{-20 \pm \sqrt{(-20)^2 - 4(1)(-96)}}{2(1)}$$

$$x = \dfrac{-20 \pm \sqrt{784}}{2}$$

$$x = \dfrac{-20 \pm 28}{2}$$

x = 8, or x = 12

Solving Quadratic Equations Exercises

4

$$x^2 + 2x - 35 = 0$$

For this quadratic equation, a = 1, b = 2, and c = -35.
Then, we simply plug these values into the quadratic formula: $x = \dfrac{-b \pm \sqrt{b^2 - 4ac}}{2a}$

$$x = \dfrac{-2 \pm \sqrt{2^2 - 4(1)(-35)}}{2(1)}$$

$$x = \dfrac{-2 \pm \sqrt{144}}{2}$$

$$x = \dfrac{-2 \pm 12}{2}$$

x = 5, or x = -7

5

$$x^2 - 5x - 24 = 0$$

For this quadratic equation, a = 1, b = -5, and c = -24.
Then, we simply plug these values into the quadratic formula: $x = \dfrac{-b \pm \sqrt{b^2 - 4ac}}{2a}$

$$x = \dfrac{-(-5) \pm \sqrt{(-5)^2 - 4(1)(-24)}}{2(1)}$$

$$x = \dfrac{5 \pm \sqrt{144}}{2}$$

$$x = \dfrac{5 \pm 11}{2}$$

x = -8, or x = -3

6

$$x^2 + 3x - 54 = 0$$

For this quadratic equation, a = 1, b = 3, and c = -54.
Then, we simply plug these values into the quadratic formula: $x = \dfrac{-b \pm \sqrt{b^2 - 4ac}}{2a}$

$$x = \dfrac{-3 \pm \sqrt{3^2 - 4(1)(-54)}}{2(1)}$$

$$x = \dfrac{-2 \pm \sqrt{225}}{2}$$

$$x = \dfrac{-2 \pm 15}{2}$$

x = 6, or x = -9

Solving Quadratic Equations Exercises

7)
$$x^2 - 16x + 63 = 0$$
For this quadratic equation, a = 1, b = -16, and c = 63.
Then, we simply plug these values into the quadratic formula: $x = \frac{-b \pm \sqrt{b^2 - 4ac}}{2a}$

$$x = \frac{-(-16) \pm \sqrt{(-16)^2 - 4(1)(63)}}{2(1)}$$

$$x = \frac{16 \pm \sqrt{256 - 252}}{2}$$

$$x = \frac{16 \pm 2}{2}$$

x = 9, or x = 7

8)
$$x^2 - 9x + 18 = 0$$
For this quadratic equation, a = 1, b = -9, and c = 18.
Then, we simply plug these values into the quadratic formula: $x = \frac{-b \pm \sqrt{b^2 - 4ac}}{2a}$

$$x = \frac{-(-9) \pm \sqrt{(-9)^2 - 4(1)(18)}}{2(1)}$$

$$x = \frac{9 \pm \sqrt{81 - 72}}{2}$$

$$x = \frac{9 \pm \sqrt{9}}{2}$$

$$x = \frac{9 \pm 3}{2}$$

x = 6, or x = 3

9)
$$x^2 - 2x - 24 = 0$$
For this quadratic equation, a = 1, b = -2, and c = -24.
Then, we simply plug these values into the quadratic formula: $x = \frac{-b \pm \sqrt{b^2 - 4ac}}{2a}$

$$x = \frac{-(-2) \pm \sqrt{(-2)^2 - 4(1)(-24)}}{2(1)}$$

$$x = \frac{2 \pm \sqrt{4 - (-96)}}{2}$$

$$x = \frac{2 \pm \sqrt{100}}{2}$$

$$x = \frac{2 \pm 10}{2}$$

x = 6, or x = -4

Solving Quadratic Equations Exercises

10

$$x^2 + 8x + 12 = 0$$

For this quadratic equation, a = 1, b = 8, and c = 12.
Then, we simply plug these values into the quadratic formula: $x = \dfrac{-b \pm \sqrt{b^2 - 4ac}}{2a}$

$$x = \dfrac{-8 \pm \sqrt{8^2 - 4(1)(12)}}{2(1)}$$

$$x = \dfrac{-8 \pm \sqrt{16}}{2}$$

$$x = \dfrac{-8 \pm 4}{2}$$

x = -6, or x = -2

11

$$x^2 - 13x + 42 = 0$$

For this quadratic equation, a = 1, b = -13, and c = 42.
Then, we simply plug these values into the quadratic formula: $x = \dfrac{-b \pm \sqrt{b^2 - 4ac}}{2a}$

$$x = \dfrac{-(-13) \pm \sqrt{(-13)^2 - 4(1)(42)}}{2(1)}$$

$$x = \dfrac{13 \pm \sqrt{169 - 168}}{2}$$

$$x = \dfrac{13 \pm \sqrt{1}}{2}$$

$$x = \dfrac{13 \pm 1}{2}$$

x = 7, or x = 4

12

$$x^2 - 7x - 60 = 0$$

For this quadratic equation, a = 1, b = -7, and c = -60.
Then, we simply plug these values into the quadratic formula: $x = \dfrac{-b \pm \sqrt{b^2 - 4ac}}{2a}$

$$x = \dfrac{-(-7) \pm \sqrt{(-7)^2 - 4(1)(-60)}}{2(1)}$$

$$x = \dfrac{7 \pm \sqrt{49 - (-240)}}{2}$$

$$x = \dfrac{7 \pm \sqrt{289}}{2}$$

$$x = \dfrac{7 \pm 17}{2}$$

x = -5, or x = 12

Solving Quadratic Equations Exercises

13)
$$8x^2 - 32x - 40 = 0$$
For this quadratic equation, a = 8, b = -32, and c = -40.
Then, we simply plug these values into the quadratic formula: $x = \frac{-b \pm \sqrt{b^2 - 4ac}}{2a}$

$$x = \frac{-(-32) \pm \sqrt{(-32)^2 - 4(8)(40)}}{2(8)}$$

$$x = \frac{32 \pm \sqrt{1024 - (-1280)}}{16}$$

$$x = \frac{32 \pm \sqrt{2304}}{16}$$

$$x = \frac{32 \pm 48}{16}$$

$$x = -1, \text{ or } x = 5$$

14)
$$18x^2 + 72x + 72 = 0$$
For this quadratic equation, a = 18, b = 72, and c = 72.
Then, we simply plug these values into the quadratic formula: $x = \frac{-b \pm \sqrt{b^2 - 4ac}}{2a}$

$$x = \frac{-72 \pm \sqrt{72^2 - 4(18)(72)}}{2(18)}$$

$$x = \frac{-72 \pm \sqrt{5184 - (5184)}}{36}$$

$$x = \frac{-72 \pm \sqrt{0}}{36}$$

$$x = \frac{-72 \pm 0}{36}$$

$$x = -2$$

15)
$$12x^2 - 12x - 72 = 0$$
For this quadratic equation, a = 1, b = -12, and c = -72.
Then, we simply plug these values into the quadratic formula: $x = \frac{-b \pm \sqrt{b^2 - 4ac}}{2a}$

$$x = \frac{-(-12) \pm \sqrt{(-12)^2 - 4(12)(-72)}}{2(12)}$$

$$x = \frac{12 \pm \sqrt{144 - (-3456)}}{24}$$

$$x = \frac{12 \pm \sqrt{3600}}{24}$$

$$x = \frac{12 \pm 60}{24}$$

$$x = -2, \text{ or } x = 3$$

Completing the Square

1) $x^2 - 8x - 20 = 0$. Add the c term to the other side of the equation: $x^2 - 8x = 20$
Divide the coefficient of b by 2, square the result, and add this quantity to both sides of the equation $(\frac{b}{2})^2$
$x^2 - 8x + 16 = 20 + 16$
Factor the trinomial on the left side:
$(x - 41)^2 = 36$
Take the square root of both sides:
$x - 4 = \pm\sqrt{36}$
$x = 4 + 6 = 10$, and $x = 4 - 6 = -2$

2) $x^2 + 18x + 56 = 0$. Add the c term to the other side of the equation: $x^2 + 18x = -56$
Divide the coefficient of b by 2, square the result, and add this quantity to both sides of the equation $(\frac{b}{2})^2$
$x^2 + 18x + 81 = -56 + 81$
Factor the trinomial on the left side:
$(x + 9)^2 = 25$
Take the square root of both sides:
$x + 9 = \pm\sqrt{25}$
$x = -9 + 5 = -4$, and $x = -9 - 5 = -14$

3) $x^2 - 6x - 55 = 0$. Add the c term to the other side of the equation: $x^2 - 6x = 55$
Divide the coefficient of b by 2, square the result, and add this quantity to both sides of the equation $(\frac{b}{2})^2$
$x^2 - 6x + 9 = 55 + 9$
Factor the trinomial on the left side:
$(x + 3)^2 = 64$
Take the square root of both sides:
$x + 3 = \pm\sqrt{64}$
$x = -3 + 8 = 5$, and $x = -3 - 8 = -11$

4) $x^2 + 12x - 64 = 0$. Add the c term to the other side of the equation: $x^2 + 12x = 64$
Divide the coefficient of b by 2, square the result, and add this quantity to both sides of the equation $(\frac{b}{2})^2$
$x^2 + 12x + 36 = 64 + 36$
Factor the trinomial on the left side:
$(x + 6)^2 = 100$
Take the square root of both sides:
$x + 6 = \pm\sqrt{100}$
$x = -6 + 10 = 4$, and $x = -6 - 10 = -16$

Completing the Square

5

$x^2 - 2x - 15 = 0$
Add the c term to the other side of the equation: $x^2 - 2x = 15$
Divide the coefficient of b by 2, square the result, and add this quantity to both sides of the equation $(\frac{b}{2})^2$
$x^2 - 2x + 1 = 15 + 1$
Factor the trinomial on the left side:
$(x - 1)^2 = 16$
Take the square root of both sides:
$x - 1 = \pm\sqrt{16}$
$x = 1 + 4 = 5$, and $x = 1 - 4 = -3$

6

$x^2 + 2x = 35$
Divide the coefficient of b by 2, square the result, and add this quantity to both sides of the equation $(\frac{b}{2})^2$
$x^2 - 2x + 1 = 35 + 1$
Factor the trinomial on the left side:
$(x + 1)^2 = 36$
Take the square root of both sides:
$x + 1 = \pm\sqrt{36}$
$x = -1 + 6 = 5$, and $x = -1 - 6 = -7$

7

$x^2 + 2x - 8 = 0$
Add the c term to the other side of the equation: $x^2 + 2x = 9$
Divide the coefficient of b by 2, square the result, and add this quantity to both sides of the equation $(\frac{b}{2})^2$
$x^2 - 2x + 1 = 9$
Factor the trinomial on the left side:
$(x + 1)^2 = 9$
Take the square root of both sides:
$x + 1 = \pm\sqrt{9}$
$x = -1 + 3 = 2$, and $x = -1 - 3 = -4$

8

$x^2 + 10x + 7 = -2$
Add the c term to the other side of the equation: $x^2 + 10x = -9$
Divide the coefficient of b by 2, square the result, and add this quantity to both sides of the equation $(\frac{b}{2})^2$
$x^2 + 10x + 25 = -9 + 25$
Factor the trinomial on the left side:
$(x + 5)^2 = 16$
Take the square root of both sides:
$x + 5 = \pm\sqrt{16}$
$x = -5 + 4 = -1$, and $x = -5 - 4 = -9$

Completing the Square

9) $x^2 + 8x + 7 = 0$. Add the c term to the other side of the equation: $x^2 + 8x = -7$
Divide the coefficient of b by 2, square the result, and add this quantity to both sides of the equation $(\frac{b}{2})^2$
$$x^2 + 8x + 16 = -7 + 16$$
Factor the trinomial on the left side:
$$(x + 4)^2 = 9$$
Take the square root of both sides:
$$x + 4 = \pm\sqrt{9}$$
$$x = -4 + 3 = -1, \text{ and } x = -4 - 3 = -7$$

10) $x^2 - 19x + 90 = 0$. Add the c term to the other side of the equation: $x^2 - 19x = -90$
Divide the coefficient of b by 2, square the result, and add this quantity to both sides of the equation $(\frac{b}{2})^2$
$$x^2 - 19x + \frac{361}{4} = -90 + \frac{361}{4}$$
Find a common denominator on the right side so that addition can be completed:
$$x^2 - 19x + \frac{361}{4} = -\frac{360}{4} + \frac{361}{4}$$
Add:
$$x^2 - 19x + \frac{361}{4} = \frac{1}{4}$$
Factor the trinomial on the left side:
$$(x - \frac{19}{2})^2 = \frac{1}{4}$$
Take the square root of both sides:
$$(x - \frac{19}{2})^2 = \pm\frac{1}{2}$$
$$x - \frac{19}{2} + \frac{1}{2} = \frac{20}{2} = 10 \text{ and } x = \frac{19}{2} - \frac{1}{2} = \frac{18}{2} = 9$$

11) $4x^2 - 12x - 4 = 12$. Divide the equation by the coefficient of a, which is 4, to get the leading coefficient to be 1: $\frac{4x^2 - 12x - 4}{4} = \frac{12}{4} = x^2 - 3x - 1 = 3$
Add the c term to the other side of the equation: $x^2 - 3x = 4$
Divide the coefficient of b by 2, square the result, and add this quantity to both sides of the equation $(\frac{b}{2})^2$
$$x^2 - 3x + \frac{9}{4} = 4 + \frac{9}{4}$$
Find a common denominator on the right side so that addition can be completed:
$$x^2 - 3x + \frac{9}{4} = \frac{16}{4} + \frac{9}{4}$$
Add:
$$x^2 - 3x + \frac{9}{4} = \frac{25}{4}$$
Factor the trinomial on the left side:
$$(x - \frac{3}{2})^2 = \frac{25}{4}$$
Take the square root of both sides:
$$x - \frac{3}{2} = \pm\frac{5}{2} \qquad x = \frac{3}{2} + \frac{5}{2} = \frac{8}{2} = 4 \text{ and } x = \frac{3}{2} - \frac{5}{2} = -\frac{2}{2} = -1$$

Completing the Square

12) $2x^2 - 22x + 60 = 0$. Divide the equation by the coefficient of a, which is 2, to get the leading coefficient to be 1: $\dfrac{2x^2 - 22x - 60}{2} = \dfrac{0}{2} = x^2 - 11x + 30 = 0$

Add the c term to the other side of the equation: $x^2 - 11x = -30$

Divide the coefficient of b by 2, square the result, and add this quantity to both sides of the equation $(\dfrac{b}{2})^2$

$$x^2 - 11x + \dfrac{121}{4} = -30 + \dfrac{121}{4}$$

Find a common denominator on the right side so that addition can be completed:

$$x^2 - 11x + \dfrac{121}{4} = -\dfrac{120}{4} + \dfrac{121}{4}$$

Add:

$$x^2 - 3x + \dfrac{121}{4} = \dfrac{1}{4}$$

Factor the trinomial on the left side:

$$(x - \dfrac{11}{2})^2 = \dfrac{1}{4}$$

Take the square root of both sides:

$$x - \dfrac{11}{2} = \pm \dfrac{1}{2} \qquad x = \dfrac{11}{2} + \dfrac{1}{2} = \dfrac{12}{2} = 5 \text{ and } x = \dfrac{11}{2} - \dfrac{1}{2} = \dfrac{10}{2} = 5$$

13) $4x^2 - 8x - 22 = 0$. Divide the equation by the coefficient of a, which is 4, to get the leading coefficient to be 1: $\dfrac{4x^2 - 8x - 21}{4} = \dfrac{0}{4} = x^2 - 2x - \dfrac{21}{4} = 0$

Add the c term to the other side of the equation: $x^2 - 2x = \dfrac{21}{4}$

Divide the coefficient of b by 2, square the result, and add this quantity to both sides of the equation $(\dfrac{b}{2})^2$

$$x^2 + 2x + 1 = \dfrac{21}{4} + 1$$

Find a common denominator on the right side so that addition can be completed:

$$x^2 - 2x + 1 = \dfrac{21}{4} + \dfrac{4}{4}$$

Add:

$$x^2 - 2x + 1 = \dfrac{25}{4}$$

Factor the trinomial on the left side:

$$(x - 1)^2 = \dfrac{25}{4}$$

Take the square root of both sides:

$$x - 1 = \pm \dfrac{5}{2} \qquad x = 1 + \dfrac{5}{2} = \dfrac{7}{2} = 3.5 \text{ and } x = 1 - \dfrac{5}{2} = -\dfrac{3}{2} = -1.5$$

Completing the Square

14) $100x^2 - 100x - 9 = 0$. Divide the equation by the coefficient of a, which is 100, to get the leading coefficient to be 1: $\dfrac{100x^2 - 100x - 9}{100} = \dfrac{0}{100} = x^2 - x + \dfrac{9}{100} = 0$

Add the c term to the other side of the equation: $x^2 + x = \dfrac{9}{100}$

Divide the coefficient of b by 2, square the result, and add this quantity to both sides of the equation $(\tfrac{b}{2})^2$

$$x^2 - x + \dfrac{1}{4} = \dfrac{9}{100} + \dfrac{1}{4}$$

Find a common denominator on the right side so that addition can be completed:

$$x^2 - x + \dfrac{1}{4} = \dfrac{9}{100} + \dfrac{25}{100}$$

Add:

$$x^2 - x + \dfrac{1}{4} = \dfrac{36}{100}$$

Factor the trinomial on the left side:

$$(x - \dfrac{1}{2})^2 = \dfrac{36}{100}$$

Take the square root of both sides:

$x - \dfrac{1}{2} = \pm \dfrac{3}{5}$ $\quad x = \dfrac{1}{2} + \dfrac{3}{5} = \dfrac{5}{10} + \dfrac{6}{10} = \dfrac{11}{10} = 1.1$ and $x = \dfrac{1}{2} - \dfrac{3}{5} = \dfrac{5}{10} - \dfrac{6}{10} = -\dfrac{1}{10} = -0.1$

---

15) $12x^2 - 24x + 6 = -4$. Divide the equation by the coefficient of a, which is 12, to get the leading coefficient to be 1: $\dfrac{12x^2 - 24x + 6}{12} = -\dfrac{4}{12} = x^2 - 2x + \dfrac{6}{12} = -\dfrac{4}{12}$

Add the c term to the other side of the equation: $x^2 - 2x = -\dfrac{4}{12} - \dfrac{6}{12}$

Simplify:

$$x^2 - 2x = -\dfrac{5}{6}$$

Divide the coefficient of b by 2, square the result, and add this quantity to both sides of the equation $(\tfrac{b}{2})^2$

$$x^2 - 2x + 1 = -\dfrac{5}{6} + 1$$

Simplify:

$$x^2 - 2x + 1 = \dfrac{1}{6}$$

Factor the trinomial on the left side:

$$(x - 1)^2 = \dfrac{1}{6}$$

Take the square root of both sides:

$x - 1 = \pm 0.408$ $\quad$ $x = 1 + 0.408 = 1.408$, and $x = 1 - 0.408 = 0.592$

Solving Systems of Equations with Elimination

1

$$x + 3y = 18;\ -x - 4y = -25$$

We can add these two equations to cancel the x term:

$$x + 3y = 18$$
$$+\ -x - 4y = -25$$

$$-y = -7$$
$$y = 7$$

Next, we can plug in our calculated value for y into either equation to find x:

$$x + 3y = 18$$
$$x + 3(7) = 18$$
$$x = -3$$

Therefore, the solution to the system is (-3, 7).

2

$$y = -\frac{3}{2}x - 7;\ y = \frac{1}{2}x + 5$$

Because both of these equations are equal to y, we can subtract one from the other to eliminate y:

$$y = \frac{1}{2}x + 5$$
$$-\ y = -\frac{3}{2}x - 7$$

$$0 = \frac{4}{2}x + 12$$

Simplify:

$$-12 = 2x$$
$$x = -6$$

Next, we can plug in our calculated value for x into either equation to find y:

$$y = \frac{1}{2}x + 5$$
$$y = \frac{1}{2}(-6) + 5$$
$$y = -3 + 5$$
$$y = 2$$

Therefore, the solution to the system is (-6, 2).

Solving Systems of Equations with Elimination

3

$$y = \frac{1}{4}x - 2; \quad y = -3x - 15$$

In this system, it is easiest to start by multiplying the first equation by 4 to get rid of the fraction:

$$4(y = \frac{1}{4}x - 2)$$
$$4y = x - 8$$

Now, we can multiply our result by 3 so that when we add the two equations, the x terms will cancel out:

$$3(4y = x - 8)$$
$$12y = 3x - 24$$

Add the two equations:

$$12y = 3x - 24$$
$$+ \quad y = -3x - 15$$

$$13y = -39$$
$$y = -3$$

Next, we can plug in our calculated value for y into either equation to find x:

$$y = -3x - 15$$
$$-3 = -3x - 15$$
$$3x = -12$$
$$x = -4$$

Therefore, the solution to the system is (-4, -3).

4

$$3x + y = -21; \quad x + y = -5$$

Recall that solving systems of linear equations via the elimination method relies on the addition property of equality, which states that states that the same value can be added or subtracted to both sides of an equation and equality is maintained. Therefore, since x + y= -5, what's on the left (x+y) equals what is on the right (-5), so we can subtract the x + y to one side of our other equation and the -5 to the other, and we haven't changed the truth of the equality statement of our other equation.

$$3x + y = -21$$
$$-(x + y = -5)$$

$$2x + 0 = -16$$

Now, we can easily solve for x by dividing both sides by 2:

$$2x + 0 = -16$$
$$x = -8$$

Once we have our value for x, we can plug it in to either of the initial equations to solve for y:

$$x + y = -5$$
$$-8 + y = -5; \quad y = 3$$

Therefore, the solution to the system is (-5, 3).

Solving Systems of Equations with Elimination

5

$-5x - 7y = 24$; $10x + 7y = 1$

We can add these two equations to cancel out the y term:

$$-5x - 7y = 24$$
$$+\ 10x + 7y = 1$$

$$5x = 25$$

Solving for x yields $x = 5$

Next, we can plug in our calculated value for x into either equation to find y:

$$10x + 7y = 1$$
$$10(5) + 7y = 1$$
$$50 + 7y = 1$$
$$7y = -49$$
$$y = -7$$

Therefore, the solution to the system is (5, -7).

6

$5x - 2y = 18$; $-2x - y = -9$

At first glance, it can be determined that these equations can't be simply added or subtracted to eliminate a variable. However, just as the addition property of equality can be used with the elimination method, so too can the multiplication property of equality, which states that the same non-zero real number can be multiplied by both sides of an equation without altering the truth of the equality statement of the original equation. Because division is the same as multiplying times a reciprocal, an equation can be divided by the same number on both sides as well. Therefore, we can multiply the second equation by -2 and then add the result to the first equation to cancel the y term:

$$-2(-2x - y = -9)$$
$$= 4x + 2y = 18$$

Now, we will add this result to the first equation, as doing so will cancel out the y term:

$$4x + 2y = 18$$
$$+\ 5x - 2y = 18$$

$$9x = 36$$
$$x = 4$$

Next, we can plug in our calculated value for x into either equation to find y:

$$5x - 2y = 18$$
$$5(4) - 2y = 18$$
$$20 - 2y = 18$$
$$2y = 2$$
$$y = 1$$

Therefore, the solution to the system is (4, 1).

Solving Systems of Equations with Elimination

7

$$y = -\frac{2}{7}x - 2; \quad y = -\frac{1}{7}x - 4$$

Here, we can multiply both equations by 7 to get rid of the fractions:

$$7(y = -\frac{2}{7}x - 2) = 7y = -2x - 14$$
$$7(y = -\frac{1}{7}x - 4) = 7y = -x - 28$$

Next, we can subtract the second from the first:

$$7y = -2x - 14$$
$$-\quad 7y = -x - 28$$
$$\overline{}$$
$$0 = -x + 14$$
$$x = 14$$

Next, we can plug in our calculated value for x into either equation to find y:

$$y = -\frac{1}{7}x - 4$$
$$y = -\frac{1}{7}(14) - 4$$
$$y = -2 - 4$$
$$y = -6$$

Therefore, the solution to the system is (14, -6).

8

$$3x + 3y = -3; \quad -3x - 4y = 2$$

These equations can be added to cancel the x term:

$$3x + 3y = -3$$
$$+ \; -3x - 4y = 2$$
$$\overline{}$$
$$-y = -1$$
$$y = 1$$

Next, we can plug in our calculated value for y into either equation to find x:

$$3x + 3y = -3$$
$$3x + 3(1) = -3$$
$$3x = -6$$
$$x = -2$$

Therefore, the solution to the system is (-2, 1).

Solving Systems of Equations with Elimination

9

$-x + 2 = y$; $-6x - 3 = y$

Because both of these equations are equal to y, we can subtract one from the other to eliminate y:

$-x + 2 = y$
$-\ \ -6x - 3 = y$

$5x + 5 = 0$
$x = -1$

Next, we can plug in our calculated value for y into either equation to find x:

$-x + 2 = y$
$-(-1) + 2 = y$
$y = 3$

Therefore, the solution to the system is (-1, 3).

10

$4x - y = -1$; $-3x + 2y = -3$

For this system, multiplying the first equation by 2 and then adding it to the second equation will cancel out the y term:

$2(4x - y = -1)$
$8x - 2y = -2$

Now, we add the two equations:

$8x - 2y = -2$
$+\ \ -3x + 2y = -3$

$5x = -5$
$x = -1$

Next, we can plug in our calculated value for x into either equation to find y:

$4x - y = -1$
$4(-1) - y = -1$
$y = -3$

Therefore, the solution to the system is (-1, -3).

Solving Systems of Equations with Substitution

1

$2x - 3y = -1;\ -x + y = -1$

First, we need to solve one of the equations in terms of the other. That seems easier with the second equation:

$-x + y = -1$

$y = x - 1$

Now, we can substitute this equation in for y in our other equation, because it will enable us to have only one variable— x—in that equation.

$2x - 3y = -1$

$2x - 3(x - 1) = -1$

Distribute:

$2x - 3x + 3 = -1$

Next, we simplify by isolating the variable on one side of the equation and the constants on the other:

$-x = -4$

Therefore, $x = 4$.

Next, we substitute the calculated value of x into the equation we manipulated to be in terms of y:

$y = x - 1$

$y = 4 - 1 = 3$

Therefore, the solution is (4, 3).

2

$y = \frac{1}{2}x + 4;\ y = -\frac{5}{2}x + 10$

Because both of the given equations are equal to y, we can set them equal to one another to solve for x. Therefore, $\frac{1}{2}x + 4 = -\frac{5}{2}x + 10$. To solve for x, we need to isolate the variable by moving all instances of x to one side of the equation. However, in this case, since we have a fraction of x on one side, it makes sense to remove the fraction by multiplying all terms by the reciprocal:

$2(\frac{1}{2}x + 4) = 2(-\frac{5}{2}x + 10)$

$x + 8 = -5x + 20$

Then, we can simplify:

$6x = 12$

Solving for x yields:

$x = 2$

Next, we substitute the calculated value of x into either of the initial equations to solve for y:

$y = \frac{1}{2}(2) + 4$

$y = 1 + 4 = 5$

Therefore, the solution is (2, 5).

Solving Systems of Equations with Substitution

3)

$$-16 = 8x + y; \; -3x + y = -5$$

First, we need to solve one of the equations in terms of the other. We will use the first equation:

$$-16 = 8x + y$$
$$y = -8x - 16$$

Now, we can substitute this equation in for y in our other equation, because it will enable us to have only one variable—x—in that equation.

$$-3x + y = -5$$
$$-3x + (-8x - 16) = -5$$

Next, we simplify by isolating the variable on one side of the equation and the constants on the other:

$$11x = -11$$

Therefore, $x = -1$.

Next, we substitute the calculated value of x into the equation we manipulated to be in terms of y:

$$y = -8x - 16$$
$$y = -8(-1) - 16$$
$$y = 8 - 16, \; y = -8$$

Therefore, the solution is $(-1, -8)$.

4)

$$y = 4x - 10; \; y = \frac{1}{3}x + 1$$

Because both of the given equations are equal to y, we can set them equal to one another to solve for x. Therefore, $4x - 10 = \frac{1}{3}x + 1$. To solve for x, we need to isolate the variable by moving all instances of x to one side of the equation. However, in this case, since we have a fraction of x on one side, it makes sense to remove the fraction by multiplying all terms by the reciprocal:

$$3(4x - 10) = 3(\frac{1}{3}x + 1)$$
$$12x - 30 = x + 3$$

Then, we can simplify:

$$11x = 33$$

Solving for x yields:

$$x = 3$$

Next, we substitute the calculated value of x into either of the initial equations to solve for y:

$$y = 4(3) - 10$$
$$y = 12 - 10 = 2$$

For completeness, it can be seen that the other equation would yield the same value for y:

$$y = \frac{1}{3}(3) + 1$$
$$y = 1 + 1 = 2$$

Therefore, the solution is $(3, 2)$.

Solving Systems of Equations with Substitution

5

$$y = 4x + 5;\ y = -\frac{1}{3}x - 8$$

Because both of the given equations are equal to y, we can set them equal to one another to solve for x. Therefore, $4x + 5 = -\frac{1}{3}x - 8$. To solve for x, we need to isolate the variable by moving all instances of x to one side of the equation. However, in this case, since we have a fraction of x on one side, it makes sense to remove the fraction by multiplying all terms by the reciprocal:

$$3(4x + 5) = 3(-\frac{1}{3}x - 8)$$
$$12x + 15 = -x - 24$$

Then, we can simplify:

$$13x = -39$$

Solving for x yields:

$$x = -3$$

Next, we substitute the calculated value of x into either of the initial equations to solve for y:

$$y = 4(-3) + 5$$
$$y = -12 + 5 = -7$$

Therefore, the solution is (-3, -7).

6

$$y = -4x + 15;\ y = -\frac{7}{2}x + 12$$

Because both of the given equations are equal to y, we can set them equal to one another to solve for x. Therefore, $-4x + 15 = -\frac{7}{2}x + 12$. To solve for x, we need to isolate the variable by moving all instances of x to one side of the equation. However, in this case, since we have a fraction of x on one side, it makes sense to remove the fraction by multiplying all terms by the reciprocal:

$$2(-4x + 15) = 2(-\frac{7}{2}x + 12)$$
$$-8x + 30 = -7x + 24$$

Then, we can simplify:

$$6 = x$$

Next, we substitute the calculated value of x into either of the initial equations to solve for y:

$$y = -4(6) + 15$$
$$y = -24 + 15 = -9$$

Therefore, the solution is (6, -9).

Solving Systems of Equations with Substitution

7) $5x + 2y = 21; -x - y = -9$

First, we need to solve one of the equations in terms of the other. We will use the second equation:

$-x - y = -9$

$y = 9 - x$

Now, we can substitute this equation in for y in our other equation, because it will enable us to have only one variable—x—in that equation.

$5x + 2y = 21$

$5x + 2(9 - x) = 21$

Distribute:

$5x + 18 - 2x = 21$

Next, we simplify by isolating the variable on one side of the equation and the constants on the other:

$3x = 3$

Therefore, $x = 1$.

Next, we substitute the calculated value of x into the equation we manipulated to be in terms of y:

$y = 9 - x$

$y = 9 - 1; y = 8$

Therefore, the solution is (1, 8).

8) $6x - 5y = 12; 2x + y = 20$

First, we need to solve one of the equations in terms of the other. We will use the second equation to solve for y:

$2x + y = 20$

$y = 20 - 2x$

Now, we can substitute this equation in for y in our other equation, because it will enable us to have only one variable—x—in that equation:

$6x - 5y = 12$

$6x - 5(20 - 2x) = 12$

Distribute:

$6x - 100 + 10x = 12$

Next, we simplify by isolating the variable on one side of the equation and the constants on the other:

$112 = 16x$

Therefore, $x = 7$.

Next, we substitute the calculated value of x into the equation we manipulated to be in terms of y:

$y = 20 - 2x$

$y = 20 - 2(7)$

$y = 20 - 14 = 6$

Therefore, the solution is (7, 6).

Solving Systems of Equations with Substitution

9

$5 = 4x - 7y; 9x - 7y = -15$

Because both of the given equations have a -7y term, we can rearrange them and set them equal to one another:

$5 = 4x - 7y$

$5 - 4x = -7y$

And:

$9x - 7y = -15$

$-7y = -15 - 9x$

Therefore:

$5 - 4x = -15 - 9x$

Now, we can simplify:

$5x = -20, x = -4$

Now, we can substitute this equation in for x in either equation to find the value of y:

$5 = 4(-4) - 7y$

$5 = -16 - 7y$

$21 = -7y; y = -3$

Therefore, the solution is (-4, -3).

10

$y = -\frac{7}{5}x - 3; y = -\frac{4}{9}x - 3$

Because both of the given equations are equal to y, we can set them equal to one another to solve for x. Therefore, $-\frac{7}{5}x - 3 = -\frac{4}{9}x - 3$. To solve for x, we need to isolate the variable by moving all instances of x to one side of the equation. However, in this case, since we have a fraction of x on both sides, we need to find a common denominator:

$-\frac{7}{5}x - 3 = -\frac{4}{9}x - 3$

$-\frac{63}{45}x - 3 = -\frac{20}{45}x - 3$

Then, we can simplify by moving the x terms to one side of the equation and the constants to the other:

$\frac{43}{45}x = 0$

Then, we can simplify by solving for x:

$x = 0$

Next, we substitute the calculated value of x into either of the initial equations to solve for y, which is -3.

Therefore, the solution is (0, -3).

Functions

1) $f(x) = x^2 - 3$

x	f(x)
-2	1
-1	-2
0	-3
1	-2
2	1

2) $f(x) = 2x + 4$

x	f(x)
-2	0
-1	2
1	6
2	8
5	14

3) $f(x) = 5x^2 - 1$

x	f(x)
-3	44
-1	4
2	19
3	44
5	124

4) $f(x) = 8 - 5x$

x	f(x)
-6	38
-3	23
0	8
3	-7
5	-17

5) $f(x) = -3x + 6$

x	f(x)
-5	21
-1	9
0	6
5	-9
7	-15

Functions

1 $f(x) = x^2 - 2$

x	f(x)
-4	-18
-3	-11
0	-2
1	-3
2	-6

2 $f(x) = 3 - 2x$

x	f(x)
-3	9
-1	5
0	3
2	-1
4	-5

3 $f(x) = 2x^3$

x	f(x)
-2	-16
-1	-2
1	2
2	16
3	54

4 $f(x) = \frac{x}{3}$

x	f(x)
-6	-2
-3	-1
3	1
6	2
9	3

5 $f(x) = \frac{x^3}{2}$

x	f(x)
-2	-4
1	0.5
2	4
3	13.5
4	32

Algebraic Word Problems

1   **$0.45**: To solve this problem, list the givens:
Store coffee = $1.23/lbs
Local roaster coffee = $1.98/1.5 lbs
Calculate the cost for 5 lbs. of store brand.
$$\frac{\$1.23}{1 \text{lbs}} \times 5 \text{ lbs} = \$6.15$$
Calculate the cost for 5 lbs. of the local roaster.
$$\frac{\$1.98}{1.5 \text{lbs}} \times 5 \text{ lbs} = \$6.60$$
Subtract to find the difference in price for 5 lbs.

$6.60
-$6.15
──────
$0.45

2   **$3,325**: List the givens.
1,800 ft. = $2,000
Cost after 1,800 ft. = $1.00/ft.
Find how many feet left after the first 1,800 ft.

3,125 ft.
-1,800 ft.
──────
1,325 ft.

Calculate the cost for the feet over 1,800 ft.
$$1,325 \text{ ft.} \times \frac{\$1.00}{1 \text{ft}} = \$1,325$$
Total for entire cost.
$2,000 + $1,325 = $3,325

3   **18**: If Ray will be 25 in three years, then he is currently 22. The problem states that Lisa is 13 years younger than Ray, so she must be 9. Sam's age is twice that, which means that the correct answer is 18.

4   **35 feet**: Denote the width as w and the length as l. Then, l = 3w + 5. The perimeter is 2w + 2l = 90. Substituting the first expression for l into the second equation yields 2(3w + 5) + 2w = 90, or 8w = 80, so w = 10. Putting this into the first equation, it yields l = 3(10) + 5 = 35 feet.

5   **6**: This problem involved setting up an algebraic equation to solve for x, or the number of flower trays Carly purchased. The equation is as follows:
6x + 8x = 84
So,
14x = 84
Then divide each side by 14 to solve for x:
$$x = \frac{84}{14} = 6 \text{ trays}$$

6   **8**: Let a be the number of apples and b the number of bananas. Then, the total cost is 2a + 3b = 22, and it also known that a + b = 10. Using the knowledge of systems of equations, cancel the b variables by multiplying the second equation by -3. This makes the equation -3a - 3b = -30. Adding this to the first equation, the b values cancel to get -a = -8, which simplifies to a = 8.

7   **4**: Let r be the number of red cans and b be the number of blue cans. One equation is r + b = 10. The total price is $16, and the prices for each can means 1r + 2b = 16. Multiplying the first equation on both sides by -1 results in -r - b = -10. Add this equation to the second equation, leaving b = 6. So, she bought 6 blue cans. From the first equation, this means r = 4; thus, she bought 4 red cans.

Algebraic Word Problems

8 **20 yo-yos**: Let y be the number of initial yo-yos in his collection. We can write the following equation from the given information:
$$16 = 6 + \frac{1}{2}y$$
Then we solve for y:
$$10 = \frac{1}{2}y$$
$$y = 20 \text{ yo-yos}$$

9 **47**: We can represent the pages in the book relative to one another. The lowest page number can be p. Then, the next consecutive page is p + 1, and the third page is p + 2. Therefore, we can write the following equation using only one letter variable:
$$144 = p + (p + 1) + (p + 2)$$
Simplifying and solving yields:
$$144 = 3p + 3$$
$$141 = 3p$$
$$p = 47$$
Thus, the lowest page is 47, then the other two are 48 and 49.

10 **43 tickets**: Let t be the number of tickets she had before buying the football. Then, we can write the following equation from the given information:
$$26 = \frac{t + 9}{2}$$
Simplifying and solving yields:
$$26 = \frac{t + 9}{2}$$
$$52 = t + 9$$
$$t = 43 \text{ tickets}$$

11 **$17**: Let c be the cost of each poster. First, we can figure out how much he spent on the posters:
$$\$165 - \$29 = \$136$$
Then, we can write the following equation from the given information:
$$\$136 = 8c$$
$$c = \$17$$

12 **46 students**: Let s be the number of students per bus. First, we can figure out the number of students who rode a bus:
$$331 - 9 = 322 \text{ students}$$
Then, we can write the following equation from the given information:
$$322 = 7s$$
$$s = 46 \text{ students}$$

13 **$9**: Let t be the cost of each ticket. Thus, we can write the following equation:
$$13 = \frac{12 + 3t}{3}$$
Simplifying and solving yields:
$$39 = 12 + 3t$$
$$27 = 3t$$
$$t = \$9$$

14 **33 cups**: First, we have to determine the profit on each cup of lemonade sold. To do this, we start by determining the cost of ingredients per cup. Because each pitcher makes 8 cups, we divide the cost per pitcher by 8: $1.12/8 = $.14 per cup. Then we can determine the profit per cup: $0.75 - $0.14 = $0.61 per cup profit. Then, to determine the number of cups she needs to sell: $20.00/$0.61 = 32.8. This needs to be rounded up to 33 cups because she can only sell whole cups.

Algebraic Word Problems

15) 174 bottles: This problem requires two equations and two variables. Let's let c be the number of aluminum cans and b be the number of glass bottles. Then we can write the following:
$$29.00 = .05c + .10b$$
And: $b = \frac{3}{4}c$
Therefore, we can substitute this value of b into our first equation to eliminate one of the variables:
$$29.00 = .05c + .10(\frac{3}{4}c)$$
$$29.00 = .125c$$
$$c = 232$$
Therefore, they had 232 cans. To find the number of bottles, we can then use our second equation:
$$b = \frac{3}{4}c$$
$$b = \frac{3}{4}(232)$$
$$b = 174$$

16) 5 batches: Again, we need two variables and two equations here. Let b be the number of brownie batches and c be the number of cookie batches. Therefore, we can write the following equations from the provided information:
$$184 = 16b + 24c \quad c = b + 1$$
Then, we can substitute this value of c relative to b into our first equation to eliminate one variable:
$$184 = 16b + 24c \quad 184 = 16b + 24(b + 1)$$
Simplifying and solving yields:
$$184 = 16b + 24(b + 1)$$
$$184 = 16b + 24b + 24$$
$$160 = 40b$$
$$b = 4$$
This means that she baked 4 batches of brownies. Since she baked one more batch of cookies than brownies, she baked 5 batches of cookies.

17) 33 games: This problem can be solved easily by dividing 44 by 4 (which is 11). This means ¼ of the games, 11 games, are home games. Therefore, the remainder (44 - 11 = 33 games) are away games.

18) 9 ladybugs: This is another instance where we can write two equations and use two variables, and then rewrite one variable in terms of the other so that we can solve for one of the variables. We will define l as the number of lady bugs and s as the number of spiders. Because ladybugs are insects, they have six legs, and spiders have eight legs. Therefore, we know the total number of legs (198) is equal to six legs per ladybug times the number of ladybugs plus eight legs per spider times the number of spiders. We also know there are twice as many spiders as ladybugs. Thus, the following equations can be written in this problem:
$$198 = 6l + 8s \quad s = 2l$$
Then, we can substitute this value of s relative to l into our first equation to eliminate one variable:
$$198 = 6l + 8s \quad 198 = 6l + 8(2l)$$
Simplifying and solving yields:
$$198 = 6l + 16l$$
$$198 = 22l$$
$$l = 9$$
Therefore, she counts 9 ladybugs. For completeness, since she counts twice as many spiders as ladybugs, she counts 9 x 2 = 18 spiders. This can be checked by plugging these values into the initial equation:
$$198 = 6(9) + 8(18) \quad 198 = 54 + 144$$
$$198 = 198$$

19) 28 students: This is another relatively simple problem. If there are seven groups with four students each, there are 4 x 7 = 28 students in the class.

20) 59.5 minutes: To solve this rate problem, we first need to determine Shankar's pace per mile in his five-mile run. To do this, we divide the total time by 5 miles. 42:30 is equal to 42.5 minutes, since 30 seconds is equal to half of one minute. Thus:
$$\frac{42.5 \text{ min}}{5 \text{ miles}} = 8.5 \text{ min/mile}$$
Next, we multiply this pace by 7 to find the time it will take him to run 7 miles:
$$8.5 \frac{\text{min}}{\text{mile}} \times 7 \text{ miles} = 59.5 \text{ minutes} = 59 \text{ minutes and } 30 \text{ seconds}$$

21) 5 days: This is another instance where we can write two equations and use two variables, and then rewrite one variable in terms of the other so that we can solve for one of the variables. We will define l as the number of long days (50 minutes) and s as the number of short days of practice (30 minutes). Because long days are 50 minutes and short days are 30 minutes, we know that the total number of weekly minutes is equal to the number of long days (l) times 50 minutes per l day plus the number of short days (s) times 30 minutes per short day. We also know the sum of the number of l and s days is seven since there are seven days in a week. First, we need to convert the weekly time to minutes so that all times have the same units:
$$5 \text{ hours} \times 60 \frac{\text{min}}{\text{hour}} + 10 \text{ min} = 310 \text{ min}$$
Now we can write our two equations:
$$310 = 50l + 30s \quad 7 = l + s$$
Next, we can write the second equation in terms of one variable relative to the other: $l = 7 - s$
Now, we can substitute this value of l into our first equation so that we only have one variable to deal with:
$$310 = 50l + 30s \quad 310 = 50(7 - s) + 30s$$
Simplifying and solving yields:
$$310 = 50(7 - s) + 30s$$
$$310 = 350 - 50s + 30s$$
$$40 = 20s \quad s = 2 \text{ days}$$
Therefore, 2 of the days are her shorter 30-minute sessions, which means that 5 days are spent playing 50 minutes since there are 7 days in a week.

Algebraic Word Problems

22 **33 hours**: Since we know that her total earnings ($396) is comprised of 1/3 tips and 2/3 of her wages, we can multiply $396 by 2/3 to find the earnings from her wages alone:
$$\$396 \times \frac{2}{3} = \$264$$
Therefore, Sam's mom earned $264 from her wages. Since she makes 8 dollars an hour, we can divide this amount by 8 to find the number of hours worked:
$$\frac{264}{\$8/\text{hr}} = 33 \text{ hours}$$

23 **$324**: This problem may seem daunting at first, but we can write an equation with the information we know. She babysat 6 times. Five of those times were four hours and one was six hours. Of the five four-hour times, four were $12 per hour and one was $15 per hour because it had an extra child. Therefore, we can write and solve the following equation:
Total earnings = 4(4 × $12)+(4 × $15)+(6 × $12)
= 4($48) + $60 + $72
= $192 + $60 + $72
= $324

24 **6 boxes:** The team needs a total of $270, and each box earns them $3. Therefore, the total number of boxes needed to be sold is 270 ÷ 3, which is 90. With 15 people on the team, the total of 90 can be divided by 15, which equals 6. This means that each member of the team needs to sell 6 boxes for the team to raise enough money to buy new uniforms.

25 **30 oranges:** One apple/orange pair costs $3 total. Therefore, Jan bought 90÷3=30 total pairs, and hence, she bought 30 oranges.

26 **4:** Kristen bought four DVDs, which would cost a total of 4 × 15=$60. She spent a total of $100, so she spent $100-$60 = $40 on CDs. Since they cost $10 each, she must have purchased 40÷10=4 CDs.

27 **390:** Three girls for every two boys can be expressed as a ratio: 3:2. This can be visualized as splitting the school into 5 groups: 3 girl groups and 2 boy groups. The number of students that are in each group can be found by dividing the total number of students by 5: 650 divided by 5 equals 1 part, or 130 students per group. To find the total number of girls, the number of students per group (130) is multiplied by how the number of girl groups in the school (3). This equals 390.

28 **$62:** Kimberley worked 4.5 hours at the rate of $10/h and 1 hour at the rate of $12/h. The problem states that her pay is rounded to the nearest hour, so the 4.5 hours would round up to 5 hours at the rate of $10/h. 5 × $10 + 1 × $12 = $50 + $12 = $62.

Points, Lines, and Planes

1. U, V, W, X, and Y

2. $\overline{UV}$, $\overline{VY}$, $\overline{VX}$, $\overline{VW}$

3. U, V, and W

4. $\overleftrightarrow{XY}$ and $\overleftrightarrow{WU}$

5. Y

6. $\overrightarrow{VX}$, $\overrightarrow{VW}$, $\overrightarrow{VY}$, $\overrightarrow{VU}$

Points, Lines, and Planes

1. N or K

2. $\overline{NM}$

3. Q, P

4. $\overleftrightarrow{LP}$ and $\overleftrightarrow{KN}$

5. M

6. $\overrightarrow{QK}$, $\overrightarrow{QP}$, $\overrightarrow{QL}$, $\overrightarrow{QN}$

Points, Lines, and Planes

1. L and T

2. B

3. D or E

4. $\overleftrightarrow{AC}$

5. E with either A, B, or C

6. Line n

7. E

Points, Lines, and Planes

1. Parallel
2. Parallel
3. Perpendicular
4. Parallel
5. Intersecting
6. Parallel
7. Intersecting
8. Perpendicular
9. Perpendicular
10. Intersecting

Where is the Middle?

1

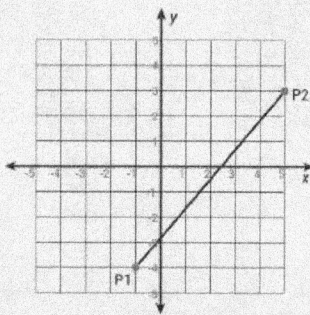

The midpoint between two points, (x_1, y_1) and (x_2, y_2), is given by taking the average of the x-coordinates and the average of the y-coordinates:
$$\left(\frac{x_1 + x_2}{2}, \frac{y_1 + y_2}{2}\right)$$
In this case, p_1 is (-1, -4) and p_2 is (5, 3). Therefore,
$$\left(\frac{x_1 + x_2}{2}, \frac{y_1 + y_2}{2}\right)$$
$$\left(\frac{-1 + 5}{2}, \frac{-4 + 3}{2}\right)$$
(0, -0.5) is the midpoint.

2

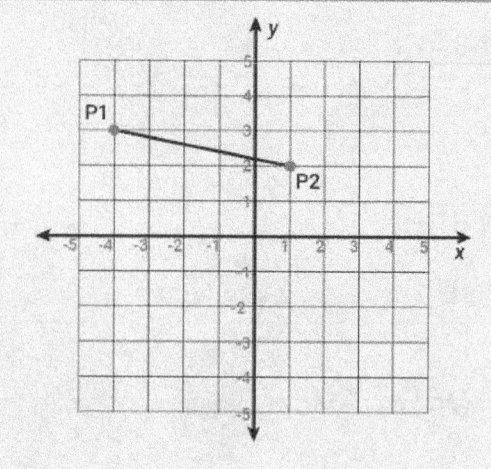

$$\left(\frac{-4 + 1}{2}, \frac{3 + 2}{2}\right)$$
(-1.5, 2.5) is the midpoint.

3

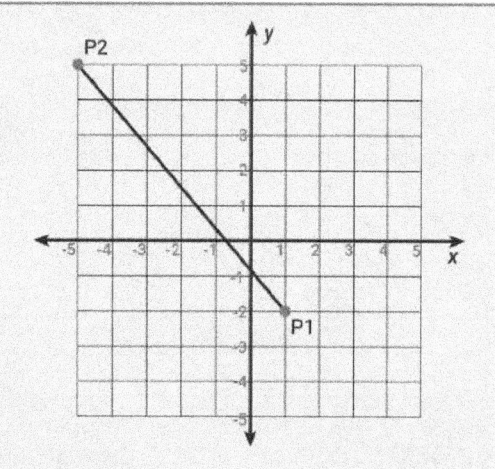

$$\left(\frac{1 + -5}{2}, \frac{-2 + 5}{2}\right)$$
(-2, 1.5) is the midpoint.

4

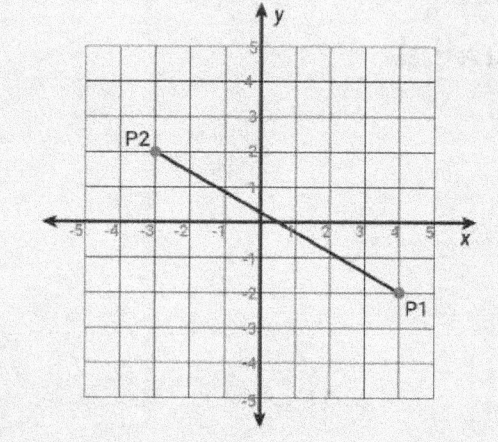

$$\left(\frac{4 + -3}{2}, \frac{-2 + 2}{2}\right)$$
(0.5, 0) is the midpoint.

Where is the Middle?

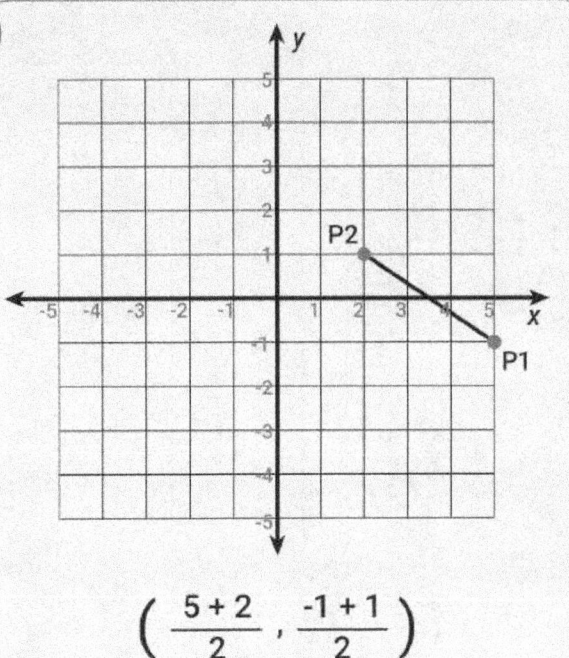

1

$\left(\dfrac{5+2}{2}, \dfrac{-1+1}{2} \right)$

(3.5, 0) is the midpoint.

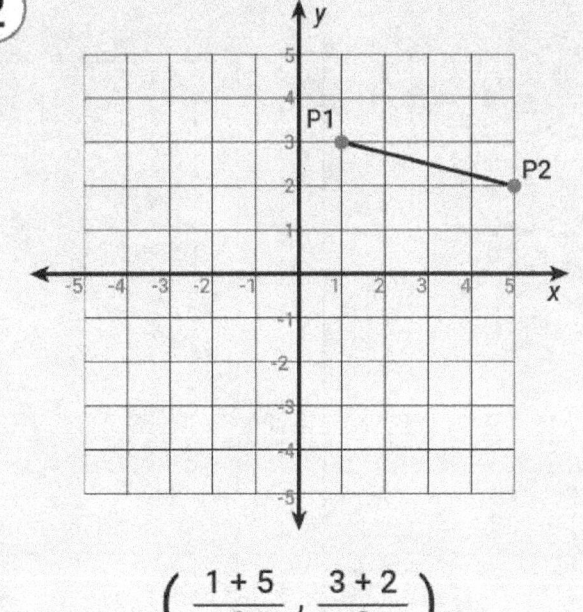

2

$\left(\dfrac{1+5}{2}, \dfrac{3+2}{2} \right)$

(3, 2.5) is the midpoint.

3

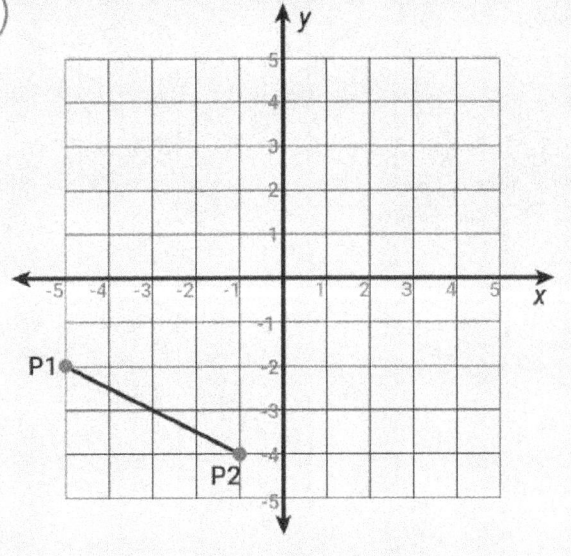

$\left(\dfrac{-5+-1}{2}, \dfrac{-2+-4}{2} \right)$

(-3, -3) is the midpoint.

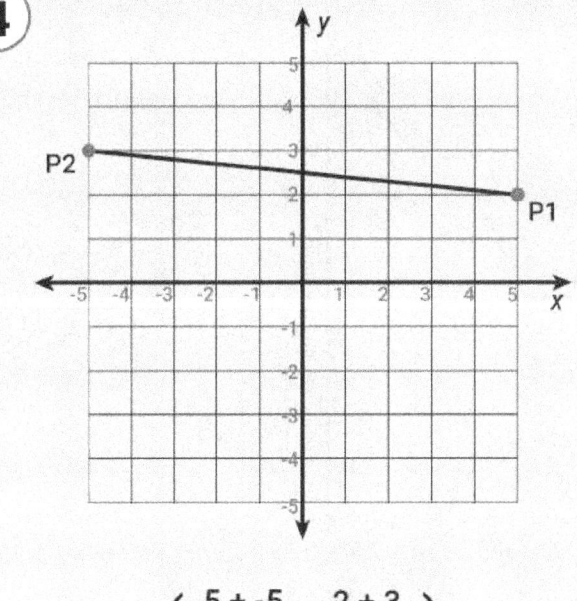

4

$\left(\dfrac{5+-5}{2}, \dfrac{2+3}{2} \right)$

(0, 2.5) is the midpoint.

Where is the Middle?

1

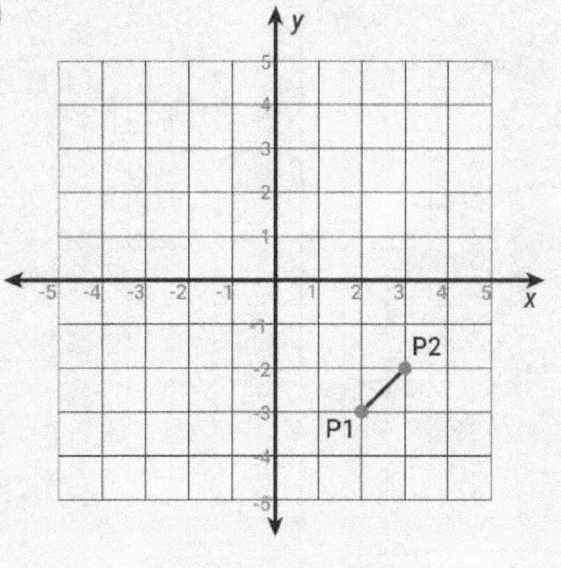

$\left(\dfrac{2 + 3}{2}, \dfrac{-3 + -2}{2} \right)$

(2.5, -2.5) is the midpoint.

2

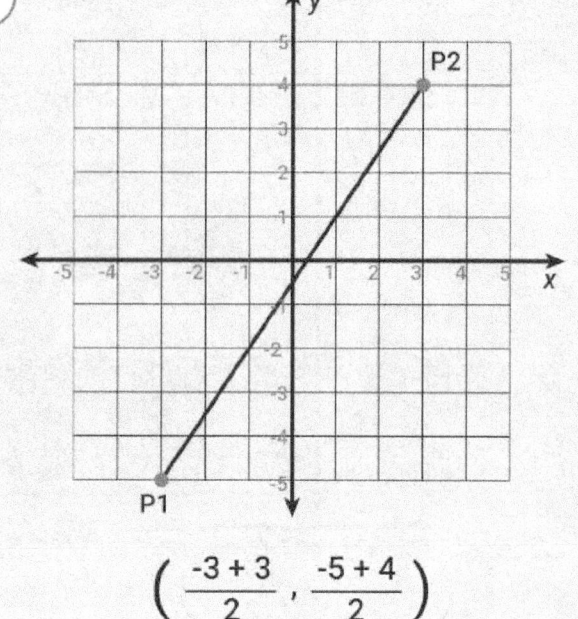

$\left(\dfrac{-3 + 3}{2}, \dfrac{-5 + 4}{2} \right)$

(0, -0.5) is the midpoint.

3

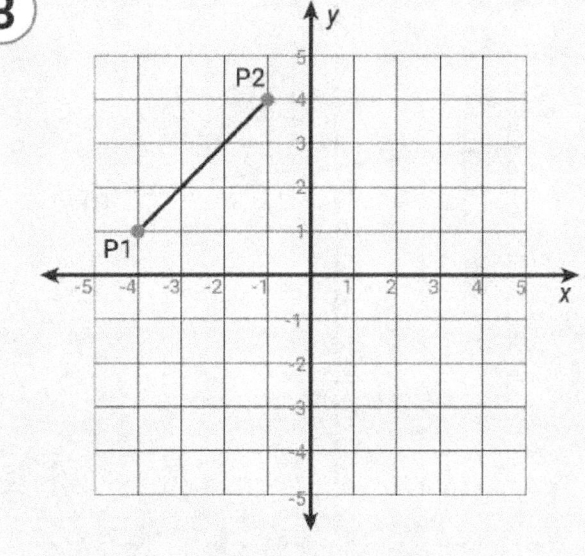

$\left(\dfrac{-4 + -1}{2}, \dfrac{1 + 4}{2} \right)$

(-2.5, 2.5) is the midpoint.

4

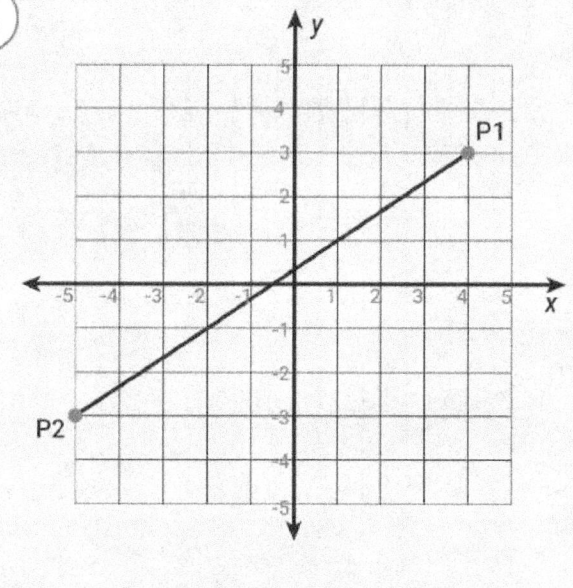

$\left(\dfrac{4 + -5}{2}, \dfrac{3 + -3}{2} \right)$

(-0.5, 0) is the midpoint.

How Far Apart?

6 A: (-2, 2); B: (4, 3)

$$d = \sqrt{(x_2 - x_1)^2 + (y_2 - y_1)^2}$$
$$d = \sqrt{(4 - -2)^2 + (3 - 2)^2}$$
$$d = \sqrt{(6)^2 + (1)^2}$$
$$d = \sqrt{37}$$
$$d = 6.1$$

2 A: (-4, 3); B: (1, -3)

$$d = \sqrt{(x_2 - x_1)^2 + (y_2 - y_1)^2}$$
$$d = \sqrt{(1 - -4)^2 + (-3 - 3)^2}$$
$$d = \sqrt{(5)^2 + (6)^2}$$
$$d = \sqrt{61}$$
$$d = 7.8$$

3 A: (-7, -3); B: (-3, -1)

$$d = \sqrt{(x_2 - x_1)^2 + (y_2 - y_1)^2}$$
$$d = \sqrt{(-3 - -7)^2 + (-1 - -3)^2}$$
$$d = \sqrt{(4)^2 + (2)^2}$$
$$d = \sqrt{20}$$
$$d = 4.5$$

4 A: (2, -1); B: (5, 4)

$$d = \sqrt{(x_2 - x_1)^2 + (y_2 - y_1)^2}$$
$$d = \sqrt{(5 - 2)^2 + (4 - -1)^2}$$
$$d = \sqrt{(3)^2 + (5)^2}$$
$$d = \sqrt{34}$$
$$d = 5.8$$

5 A: (0, 7); B: (4, 0)

$$d = \sqrt{(x_2 - x_1)^2 + (y_2 - y_1)^2}$$
$$d = \sqrt{(4 - 0)^2 + (0 - 7)^2}$$
$$d = \sqrt{(4)^2 + (-7)^2}$$
$$d = \sqrt{65}$$
$$d = 8.1$$

Determining the Measurement of Missing Angles

1. 25°

2. 30°

3. 60°

4. 21°

5. 48° + x° = 116°   x = 116 − 48   x = 68°

6. x° = 23° + 11° = 34°

7. 35° + x° = 84°   x = 84 − 35   x = 49°

8. x° = 91° + 48° = 139°

9. x° = 27° − 12° = 15°

10. x° = 125° - 92° = 33°

Using Angle Relationships to Determine the Missing Measures

1) Y = 52 degrees, x = 24 degrees

2) 95 degrees

3) ∠1 = 75, ∠2 = 105, ∠3 = 75, ∠4 = 105, ∠5 = 75, ∠6 = 105, ∠7 = 75, ∠8 = 105

4) 100 degrees

5) 65 degrees

6) 87 degrees

7) 45 degrees

8) Each of the five angles in a regular pentagon has a measure of 108 degrees because the sum of the internal angles is 180 x (n - 2). 180 x 3 = 540 degrees. 540 degrees divided by 5 angles is 108 degrees per angle. Thus, angles z and y are 180 − 108 = 72 degrees. Angle x is 180 − 72 − 72 = 36 degrees.

9) 123 degrees

10) 53 degrees

Classifying Triangles

1. Scalene, right

2. Isosceles, right

3. Scalene right

4. Isosceles, right

5. Scalene, acute

6. Scalene, right

7. Equilateral, acute

8. Scalene, right

Classifying Triangles

1. Scalene, right

2. Scalene, acute

3. Isosceles, acute

4. Isosceles, acute

5. Scalene, acute

6. Scalene, acute

7. Scalene acute

8. Equilateral, acute

Calculating Perimeter

1. 32 inches. 8 inches x 4 sides = 32 inches

2. 31.4 cm. P = 2πr = 10π cm.

3. The first triangle. The first triangle has a perimeter of 4 + 9 + 8 inches, which is 21 inches, while the perimeter of the equilateral triangle is 3 x 6 inches, which is 18 inches. Therefore, the first triangle has a larger perimeter.

4. 42 cm. A hexagon has six sides, so 6 x 7 cm = 42 cm.

5. 242 meters. The perimeter of a rectangle is P = 2l + 2w so in this case, P = 2(37) + 2(84) = 242 meters.

6. 32 cm.

7. 60 mm.

8. 20 inches.

9. 68 yards.

10. 14π or 44.00 feet.

Calculating Perimeter

1) 72 cm.

2) 50 inches.

3) 42 cm.

4) 64 feet.

5) 15 yards.

6) 64 inches.

7) 24 inches.

8) If the length is twice the width, the length is 2 x 8 = 16. 16 + 16 + (2 x 8) = 48 cm.

9) 14 + 14 + (14 / 2) = 35 inches.

10) 27 / 3 = 9 + 9 + 27 + 27 = 72 cm.

Finding the Missing Side Lengths

1) 9 inches. The perimeter of a rectangular paper is 32 inches. One side is 7 inches. That means there are 14 inches for those two sides. 32 − 14 = 18 inches left for the other two sides. 18 inches divided by two more sides is 9 inches per side.

2) 9 inches. 36 / 4 = 9 inches.

3) 5 inches. 20 / 4 = 5 inches.

4) 58 - (2 x 14) = 30 inches. 30 / 2 = 15 inches.

5) 36 - 20 = 16. 16 / 2= 8 inches.

6) 17 yards. 65 - (29 + 19) = 17 yards.

7) 37 feet. 148 / 4 = 37 feet.

8) 4 cm. 19 - (7 + 8) = 4 cm.

Finding the Missing Side Lengths

1. 48 / 6 = 8 cm.

2. 6 cm. 17 - (5 + 6) = 6 cm.

3. 12 inches. 96 / 8 = 12 inches.

4. 26 meters. 122 - (2 x 35) = 52. 52 / 2 = 26 meters.

5. 27 inches. 90 - (2 x 18) = 54. 52 / 2 = 27 inches.

6. 23 inches. 140 - (2 x 47) = 46. 46 / 2 = 23 inches.

7. 9 cm. 45 / 5 = 9 cm.

8. 12 cm. 42 − (17 + 13) = 12 cm.

Finding the Missing Side Lengths

1) 20 mm. 85 − (17 + 11 + 19 + 18) = 20 mm.

2) 30 cm. 120 / 4 = 30 cm.

3) 5 cm. 9 + 12 = 21. 26 − 21 = 5 cm.

4) 4 inches. 14, 13, 18, 22, 16. 14 + 13 + 18 + 22 + 16 = 83 in, so the remaining side is 4 inches.

5) 5 inches. 45 in − (4 + 12 + 9 + 15) = 5 inches.

6) 35 inches. 241 in − (24 + 39 + 27 + 31 + 17 + 33 + 35) = 35 inches.

7) 11 meters. The area of a rectangle is length x width so 88 / 8 = 11 meters.

8) 36 inches. The square root of 81 is 9 inches. If 9 inches is the length of one side, the length is 9 x 4 = 36 inches.

Area of Triangles

1 We are given a base (a) and height (b) for this right triangle, so we just need to plug the values into the formula for the area of a triangle to calculate the area: A = $\frac{1}{2}$bh = $\frac{1}{2}$4 × 3 = 6 yd².

2 This is a scalene triangle, but we are given the measurement of the height, so we can plug the value of the base (a) and the height into the formula to calculate the area: A = $\frac{1}{2}$bh = $\frac{1}{2}$6 × 3 = 9 cm².

3 This is a scalene triangle, but we are given the measurement of the height, so we can plug the value of the base (a) and the height into the formula to calculate the area:
A = $\frac{1}{2}$bh = $\frac{1}{2}$98 × 48 = 2352 ft².

4 We are given a base (a) and height (b) for this right triangle, so we just need to plug the values into the formula for the area of a triangle to calculate the area: A = $\frac{1}{2}$bh = $\frac{1}{2}$5 × 12 = 30 ft².

5 Because this is an isosceles triangle, we know that the two equal sides (labeled b) can be taken as the hypotenuse of a right triangle formed if we drop a median line in the triangle bisecting the base, a. Then, the hypotenuse is 65 meters. To calculate the height, we have to apply the Pythagorean Theorem using our bisected base with a resultant length of 20.5 meters. Therefore, we know that a² + b² = c², or 65² - 20.5² = height². So, 4225 - 420.25 = 3804.75, so the height is 61.6 meters. Therefore, the area A = $\frac{1}{2}$bh = $\frac{1}{2}$41 × 61.6 = 1264 m².

Area of Triangles

6) If each side is 4, the hypotenuse is 4 feet. To calculate the height, we have to apply the Pythagorean Theorem ($a^2 + b^2 = c^2$). Because it is an equilateral triangle, the height will be located along the median line of the triangle, which means the base would be bisected, with each half being 2 feet. Therefore, we know that $4^2 - 2^2 = height^2$, so $16 - 4 = 12$. Therefore, the height is 3.46 feet. Therefore, the area $A = \frac{1}{2}bh = \frac{1}{2} 4 \times 3.46 = 6.92$ ft².

7) We are given a base (a) and height (b) for this right triangle, so we just need to plug the values into the formula for the area of a triangle to calculate the area: $A = \frac{1}{2}bh = \frac{1}{2} 24 \times 7 = 84$ cm².

8) Because this is an isosceles triangle, we know that the two equal sides (labeled b) can be taken as the hypotenuse of a right triangle formed if we drop a median line in the triangle bisecting the base, a. Then, the hypotenuse is 5 yards. To calculate the height, we have to apply the Pythagorean Theorem using our bisected base with a length of 2 yards. Therefore, we know that $5^2 - 2^2 = height^2$. So, $25 - 4 = 21$; thus, the height is 4.58 yards. Therefore, the area is $A = \frac{1}{2}bh = \frac{1}{2} 4 \times 4.58 = 9.16$ yd².

9) We are given a base (a) and height (b) for this right triangle, so we just need to plug the values into the formula for the area of a triangle to calculate the area: $A = \frac{1}{2}bh = \frac{1}{2} 12 \times 16 = 96$ mm².

10) If each side is 8, the hypotenuse is 8 inches. To calculate the height, we have to apply the Pythagorean Theorem. Because it is an equilateral triangle, the height will be located along the median line of the triangle, which means the base would be bisected, with each half being 4 inches. Therefore, we know that $8^2 - 4^2 = height^2$. So, $64 - 16 = 48$ inches; thus, the height is 6.93 inches. Therefore, the area is $A = \frac{1}{2}bh = \frac{1}{2} 8 \times 6.93 = 27.72$ in².

Area of Triangles

11. We are given a base (a) and height (b) for this right triangle, so we just need to plug the values into the formula for the area of a triangle to calculate the area: $A = \frac{1}{2}bh = \frac{1}{2} 84 \times 56 = 2352$ mm²

12. Because this is an isosceles triangle, we know that the two equal sides (labeled b) can be taken as the hypotenuse of a right triangle formed if we drop a median line in the triangle bisecting the base, a. Then, the hypotenuse is 12 feet. To calculate the height, we have to apply the Pythagorean Theorem using our bisected base with a resultant length of 5 feet.
Therefore, we know that $12^2 - 5^2 =$ height². So, $144 - 25 = 119$; thus, the height is 10.91 feet.
Therefore, the area is $A = \frac{1}{2}bh = \frac{1}{2} 10 \times 10.91 = 54.55$ ft².

13. We are given a base (a) and height (b) for this right triangle, so we just need to plug the values into the formula for the area of a triangle to calculate the area: $A = \frac{1}{2}bh = \frac{1}{2} 70 \times 56 = 1960$ yd²

14. We are given a base (a) and height (b) for this right triangle, so we just need to plug the values into the formula for the area of a triangle to calculate the area: $A = \frac{1}{2}bh = \frac{1}{2} 75 \times 55 = 2062.5$ mm²

15. Because this is an isosceles triangle, we know that the two equal sides (labeled b) can be taken as the hypotenuse of a right triangle formed if we drop a median line in the triangle bisecting the base, a. Then, the hypotenuse is 72 inches. To calculate the height, we have to apply the Pythagorean Theorem using our bisected base with a resultant length of 29 inches. Therefore, we know that $a^2 + b^2 = c^2$, or $72^2 - 29^2 =$ height². So, $5184 - 841 = 4343$, so the height is 65.90 inches.
Therefore, the area $A = \frac{1}{2}bh = \frac{1}{2} 58 \times 65.90 = 1911$ in.²

Area of Mixed Shapes and Figures

1) 9 m². The formula for the area of a square is A = s², so a square with a side length of 3 meters has an area of A = s² = 3² = 9 m².

2) 55 ft². The formula for the area of a rectangle is A = lw, so a rectangle with length of 11 feet and a width of 5 feet has an area of A = lw = 11 × 5 = 55 ft².

3) 56 in². The formula for the area of a rectangle is A = lw, so a rectangle with length of 8 inches and a width of 7 inches has an area of A = lw = 8 × 7 = 56 in².

4) 36 cm². The formula for the area of a square is A = s², so a square with a side length of 6 centimeters has an area of A = s² = 6² = 36 cm².

5) 14 in². The area of a triangle is A = $\frac{1}{2}$ × b × h, so a triangle with a base of 4 in. and a height of 7 in. has an area of A = $\frac{1}{2}$ × 4 × 7 = 14 in².

6) 48 cm². The area of a triangle is A = $\frac{1}{2}$ × b × h, so a triangle with a base of 12 cm and a height of 8 cm has an area of A = $\frac{1}{2}$ × 12 × 8 = 48 cm².

7) 225 in². The formula for the area of a square is A = s², so a square with a side length of 15 inches has an area of A = s² = 15² = 225 in².

8) 60 in². The formula for the area of a rectangle is A = lw, so a rectangle with length of 12 inches and a width of 5 inches has an area of A = lw =12 × 5=60 in².

9) 254.34 cm². The area of a circle is calculated through the formula A = π × r². Therefore, a circle with a radius of 9 cm has an area of A = π × r² = A = π × 9² = 81π = (81 × 3.14) = 254.34 cm².

10) 10 cm². The area of a triangle is A = $\frac{1}{2}$ × b × h, so a triangle with a base of 4 cm and a height of 5 cm has an area of A = $\frac{1}{2}$ × 4 × 5 = 10 cm².

Area of Mixed Shapes and Figures

1) 914 yd². This compound figure has a rectangle and two semicircles. Therefore, the area is found by founding the area of a rectangle that is 30 yards by 20 yards and a circle with a radius of 10 yards. Note that the width of the rectangle is the diameter of the circle, so it is 2 x 10 or 20 yards. The area is A = lw + (π × r²) = (20 x 30) + (π × 10²) = 600 + 100π = 914 yd².

2) 12.3 in². This compound figure has a triangle and a semicircle. Therefore, the area is found by founding the area of a triangle that has a base of 4 inches and a height of 3 inches and half the area of a circle with a radius of 2 inches. Note that the base of the triangle is the diameter of the circle, so it is 2 x 2 or 4 inches.
The area is A = ($\frac{1}{2}$ × b × h) + $\frac{1}{2}$(π × r²) = ($\frac{1}{2}$ × 4 × 3) + $\frac{1}{2}$(π × 2²) = 6 + 2π = 12.3 in².

3) 688.5 cm². This compound figure has a rectangle and a semicircle. Therefore, the area is found by founding the area of a rectangle that is 21 cm. by 28 cm. and half the area of a circle with a radius of 8 cm. The area is A = lw + $\frac{1}{2}$(π × r²) = (21 x 28) + $\frac{1}{2}$(π × 8²) = 588 + 32π = 688.5 cm².

4) 61.1 ft². This compound figure has two equal triangles and a semicircle. Therefore, the area is found by founding the area of a triangles with a base of 4 feet and a height of 9 feet and doubling it (since there are two) and then adding it to half the area of a circle with a radius of 4 feet. Note that the base of the triangle is the radius of the circle because two triangles fit across the full diameter of the circle.
The area is A = 2($\frac{1}{2}$ × b × h) + $\frac{1}{2}$(π × r²) = 2($\frac{1}{2}$ × 9 × 4) + $\frac{1}{2}$(π × 4²) = 36 + 8π = 61.1 ft².

5) 370.2 m². This compound figure has a triangle and a semicircle. Therefore, the area is found by founding the area of a triangle that has a base of 18 meters and a height of 27 meters and half the area of a circle with a radius of 9 meters. Note that the base of the triangle is the diameter of the circle, so it is 2 x 9 or 18 meters.
The area is A = ($\frac{1}{2}$ × 18 × 27) + $\frac{1}{2}$(π × 9²) = ($\frac{1}{2}$ × 4 × 3) + $\frac{1}{2}$(π × 2²) = 243 + 40.5π = 370.2 m².

6) 354 m². The area of this compound figure can be found by adding the area of two rectangles within it. The first is on the left: 14 m x 15 m. The second is the smaller part on the right, which is actually a square (12 x 12 m). Therefore the area of the figure is the sum of those two components: (14 x 15) + (12 x 12) = 354 m².

Area of Mixed Shapes and Figures

1) 300 yd². The area of this compound figure can be found by adding the area of two rectangles within it. The first is on the bottom half: 10 yd x 20 yd. The second is the smaller part on the top half, which is actually a square (10 x 10 yd). Therefore, the area of the figure is the sum of those two components: (10 x 20) + (10 x 10) = 300 yd².

2) 800 cm². The area of this compound figure can be found by adding the area of two rectangles within it. The first is on the bottom half: 10 cm x 40 cm. The second is the smaller part on the top half, which is actually a square (20 x 20 cm). Therefore, the area of the figure is the sum of those two components: (10 x 40) + (20 x 20) = 800 cm².

3) 33.8 in². This compound figure has a triangle and a rectangle. Therefore, the area is found by founding the area of a triangle that has a base of 4.5 inches and a height of 3 inches and the area of a rectangle with a length of 6 inches and a width of 4.5 inches. Note that the base of the triangle is the width of the rectangle, so it is 4.5 inches.
The area is A = ($\frac{1}{2}$ × b × h) + lw = ($\frac{1}{2}$ × 4.5 × 3)+(4.5 × 6) = 6.75 + 27 = 33.8 in².

4) 262 in². This compound figure has a triangle and a semicircle. Therefore, the area is found by founding the area of a triangle that has a base of 10 inches and a height of 21 inches and half the area of a circle with a radius of 10 inches. Note that the base of the triangle is the radius of the circle, so it is 10 inches.
The area is A=($\frac{1}{2}$ × b × h) + $\frac{1}{2}$(π × r²) = ($\frac{1}{2}$ × 10 × 21) + $\frac{1}{2}$(π × 10²) = 105 + 50π = 262 in².

5) 4704 in². The area of this compound figure can be found by adding the area of two rectangles within it. The first is on the bottom half: 42 in x 70 in. The second is the smaller part on the top half, which is actually a square (42 x 42 in). Therefore the area of the figure is the sum of those two components: (42 x 70) + (42 x 42) = 4704 in².

6) 895.2 yd². This compound figure has a rectangle and a semicircle. Therefore, the area is found by founding the area of a rectangle that is 24 yd. by 32 yd. and half the area of a circle with a radius of 9 yd.
The area is A = lw + $\frac{1}{2}$(π × r²)= (24 x 32) + $\frac{1}{2}$(π × 9²) = 768 + 40.5π = 895.2 yd².

Volume

1) 1728 cm³. The equation used to calculate the volume of a cube (a × a × a) or a³. In this problem, the length of a side of the cube is 12 cm, so the volume is calculated by utilizing the formula (12 × 12 × 12) = 1728 cm³.

2) 5832 mm³. The equation used to calculate the volume of a cube (a × a × a) or a³. In this problem, the length of a side of the cube is 18 millimeters, so the volume is calculated by utilizing the formula (18 × 18 × 18)=5832 mm³.

3) 231 in³. The equation used to calculate the volume of a rectangular prism, like a cereal box, is length times width times height. Therefore, since our a = 3 in, b = 7 in, and c = 11 in, the volume is calculated by utilizing the formula 3 × 7 × 11= 231 in³.

4) 268 in³. The formula to calculate the volume of a sphere is $\frac{4}{3}\pi r^3$. Therefore, if the radius of a soccer ball is 4 inches, the volume of the sphere is calculated by utilizing the formula $\frac{4}{3}\pi(4)^3 = \frac{4}{3}(64)\pi = 85.3\pi = 268$ in³.

5) 600 yd³. The equation used to calculate the volume of a rectangular prism is length times width times height. Therefore, since our l = 25 yd, w = 12 yd, and d = 2 yd, the volume is calculated by utilizing the formula 25 × 12 × 2 = 600 yd³.

6) 2,574,466.7 m³. The formula to calculate a pyramid's volume is (L × W × H) ÷ 3. The Great Pyramid has a square base, so it's (230m × 230m × 146m) ÷ 3 = 2,574,466.7 m³. That's huge!

7) 21 in³. The formula used to calculate the volume of a cone is $\frac{1}{3}\pi r^2 h$. Therefore, the volume of the snow cone is calculated by utilizing the formula $\frac{1}{3}\pi 2^2 \times 5 = 6.667\pi$ in³. After substituting 3.14 for π, the volume is 21 in³.

8) 288 in³. The equation used to calculate the volume of a rectangular prism, like a chocolate box, is length times width times height. Therefore, since our l = 12 in, w = 8 in, and d = 3 in, the volume is calculated by utilizing the formula 12 × 8 × 3 = 288 in³.

Volume

1) 150 in³. The equation used to calculate the volume of a rectangular pyramid is $= \frac{l \times w \times h}{3}$, so plugging in our values gives the answer:
$V = \frac{5 \times 9 \times 10}{3} = 150$ in³.

2) 36,172.8 m³. The equation used to calculate the volume of a cylinder is $V = \pi r^2 h$, so plugging in our values gives the answer: $V = \pi(12)^2 80 = 11520\pi$ m³.
After substituting 3.14 for π, the volume is 36,172.8 m³.

3) 840 in³. The equation used to calculate the volume of a rectangular prism is length times width times height. Therefore, since our l = 15 in, w = 7 in, and d = 8 in, the volume is calculated by utilizing the formula 15 × 7 × 8 = 840 in³.

4) 6912 in³. The equation used to calculate the volume of a rectangular prism is length times width times height. Therefore, since our l = 24 in, w = 18 in, and d = 16 in, the volume is calculated by utilizing the formula 24 × 18 × 16 = 6912 in³.

5) 100.48 ft³. The formula used to calculate the volume of a cone is $\frac{1}{3}\pi r^2 h$. Therefore, the volume of the teepee is calculated by utilizing the formula $\frac{1}{3}\pi(4)^2 \times 6 = 32\pi$ ft³. After substituting 3.14 for π, the volume is 100.48 ft³. Keep in mind that the diameter was given (8 feet), so the radius is 4 feet.

6) 401.92 m³. The equation used to calculate the volume of a cylinder is $V = \pi r^2 h$, so plugging in our values gives the answer: $V = \pi(4)^2 8 = 128\pi$ m³. After substituting 3.14 for π, the volume is 401.92 m³.

7) 350 in³. The equation used to calculate the volume of a rectangular prism is length times width times height. Therefore, since our l = 10 in, w = 7 in, and d = 5 in, the volume is calculated by utilizing the formula 10 × 7 × 5 = 350 in³.

8) 50.24 in³. The equation used to calculate the volume of a cylinder is $V = \pi r^2 h$, so plugging in our values gives the answer: $V = \pi(2)^2 4 = 16\pi$ in³. After substituting 3.14 for π, the volume is 50.24 in³.

Volume

1) 523.3 in³. The formula to calculate the volume of a sphere is $\frac{1}{3}\pi r^3$. Therefore, if the radius of a is 4 inches, the volume of the sphere is calculated by utilizing the formula $\frac{1}{3}\pi(5)^3 = \frac{1}{3}(125)\pi = 166.67\pi = 523.3$ in³.

2) 113.04 cm³. The formula to calculate the volume of a sphere is $\frac{1}{3}\pi r^3$. Therefore, if the radius of is 3 cms (half the 6 cm. diameter), the volume of the sphere is calculated by utilizing the formula $\frac{1}{3}\pi(3)^3 = \frac{1}{3}(27)\pi = 36\pi = 113.04$ cm³.

3) 1356.48 in³. The formula used to calculate the volume of a cone is $\frac{1}{3}\pi r^2 h$. Therefore, the volume of the teepee is calculated by utilizing the formula $\frac{1}{3}\pi(6)^2 \times 36 = 432\pi$ in³. After substituting 3.14 for π, the volume is 1356.48 in³. Keep in mind that the diameter was given (12 in), so the radius is 6 in.

4) 4.2 in³. The formula to calculate the volume of a sphere is $\frac{4}{3}\pi r^3$. Therefore, if the radius of is 1 inch the volume of the sphere is calculated by utilizing the formula $\frac{4}{3}\pi(1)^3 = \frac{4}{3}\pi = 1.33\pi = 4.2$ in³.

5) The volume of a rectangular prism, like a toy chest, is calculated using the formula V = l × w × h, where l is length, w is width, and h is height. Therefore, the volume of the toy chest is V = 3 × 2 × 2 = 12 cubic feet.

6) The area of the top is 9 x 13. There are two sides that are 3 x 13 and two sides that are 3 x 9. The total area to be frosted is the sum of these: The surface area formula for a rectangular prism or a general box is SA = 2(lw + lh + wh), where l is the length, h is the height, and w is the width. Substituting in the dimensions of the cake yields SA = (9 × 13) +2(13 × 3) +2(9 × 3))= (117 + 78 + 54) = 249 square inches.

7) The formula to find the volume of a cylinder is V=$\pi r^2 h$, so for the snare drum V = 3.14 × 6² × 8 = 904.32 cubic inches.

8) The volume of a rectangular prism, like a hay bale, is calculated using the formula V=l×w×h, where l is length, w is width, and h is height. So, the volume of each hay bale 36 × 30 × 40 = 43,200 square inches. If he has 12 across and 6 high, he has 12 × 6 = 72 hay bales, so we need to multiply the volume of each hay bale by 72: 72 × 43,200 = 3,110,400 cubic inches.

Are They Similar?

1.
- 3 × 14
- 18 × 84

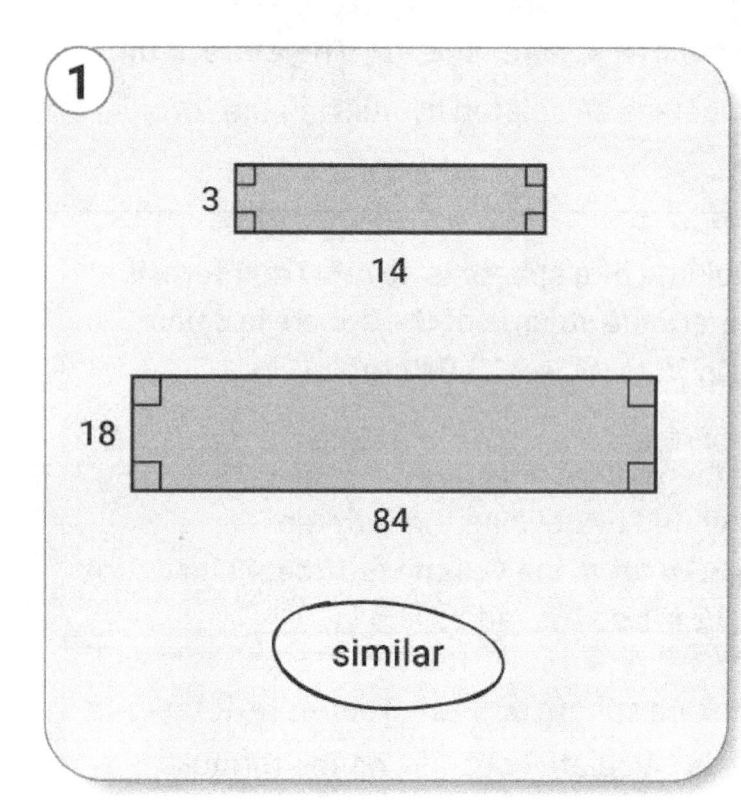

similar

2.
- 10, 30
- 25, 85

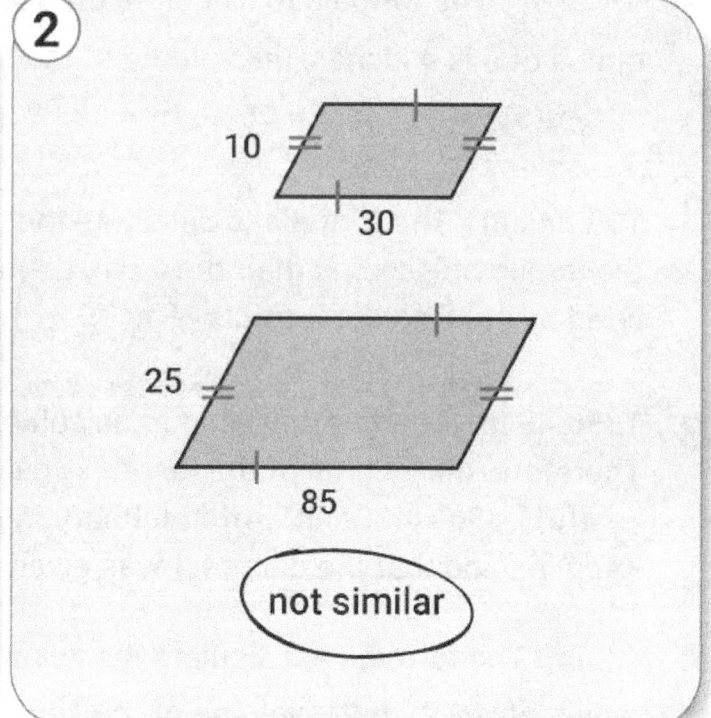

not similar

3.
- 4 × 28
- 10 × 70

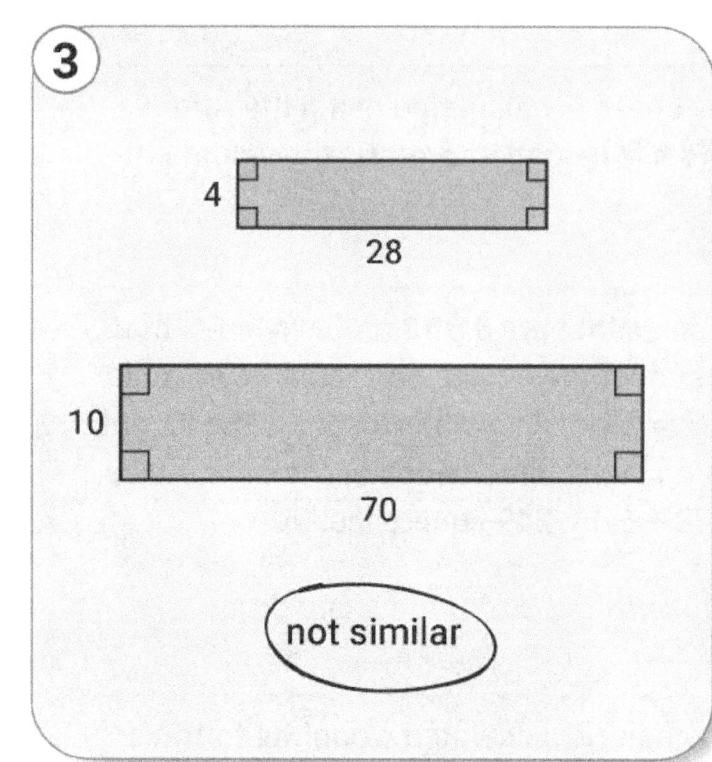

not similar

4.
- Triangle: 9, 9, 4
- Triangle: 36, 36, 12

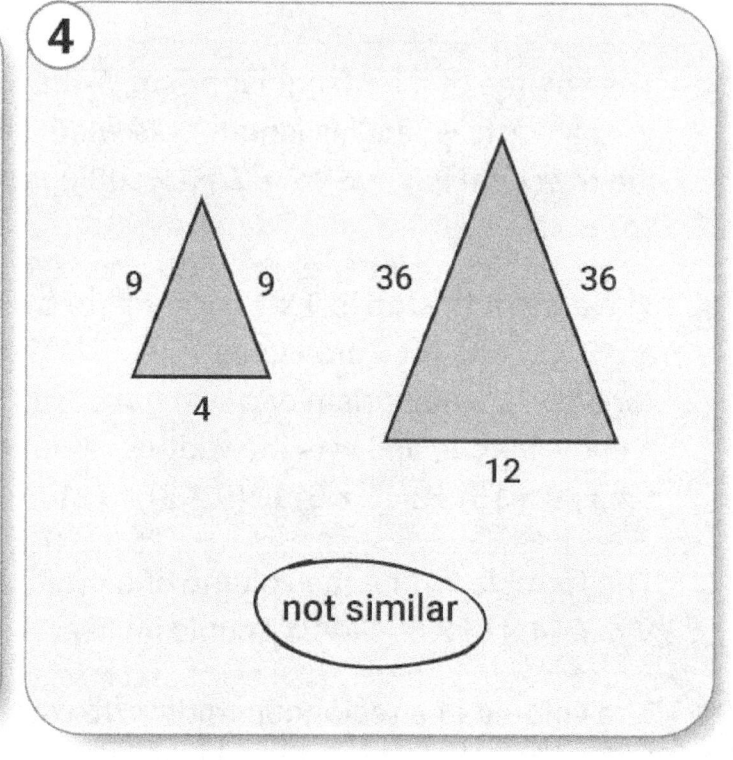

not similar

Are They Similar?

1.

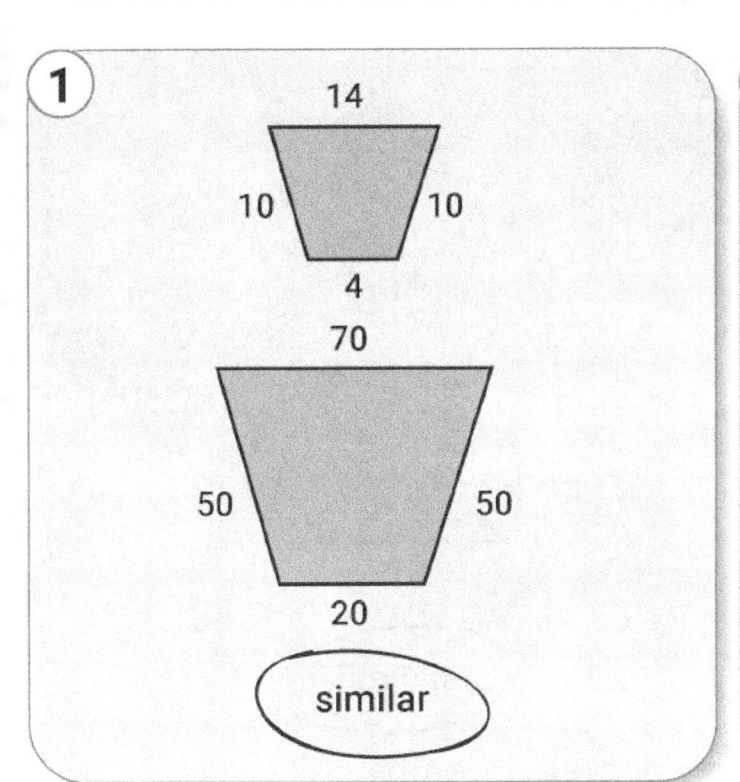

similar

2.

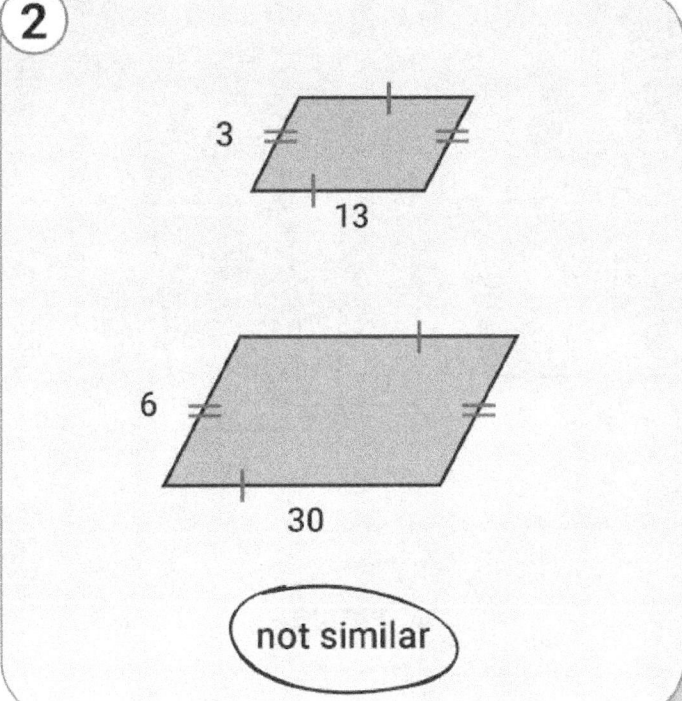

not similar

3.

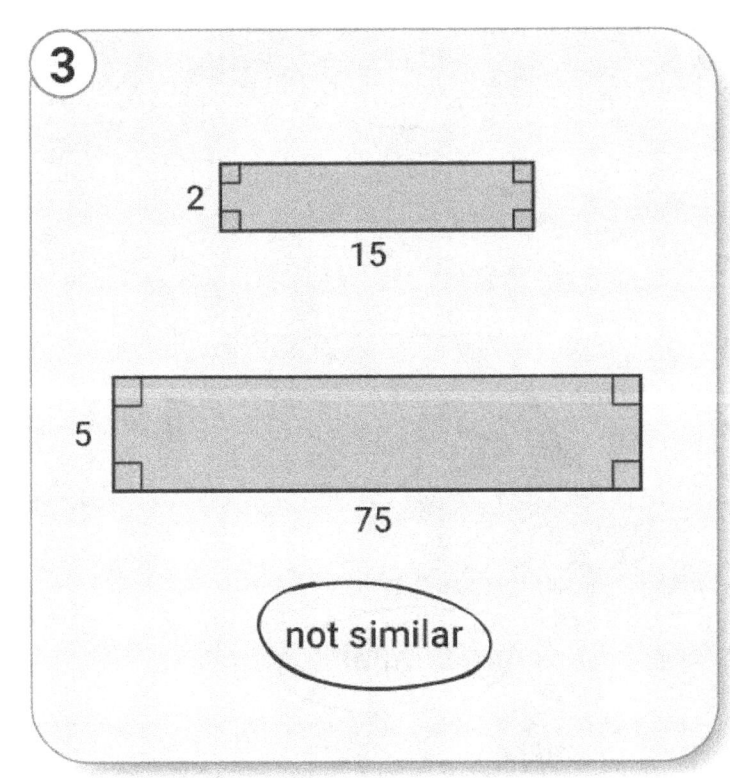

not similar

4.

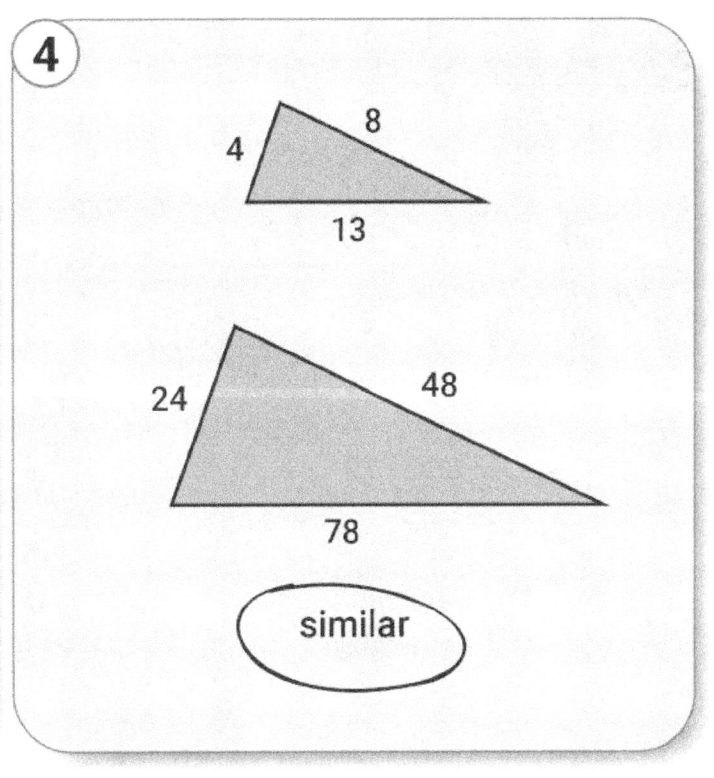

similar

Are They Similar?

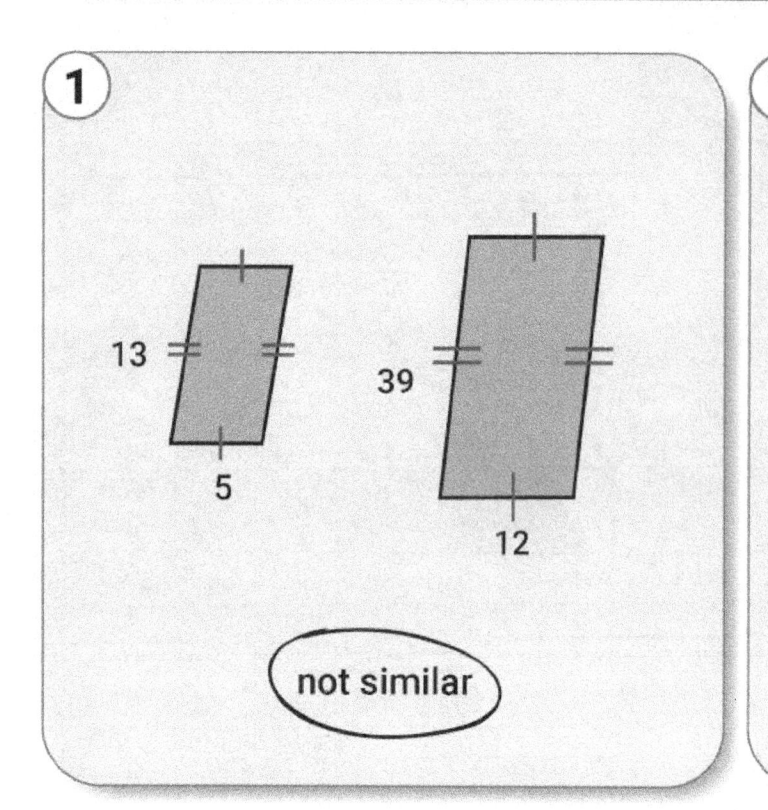

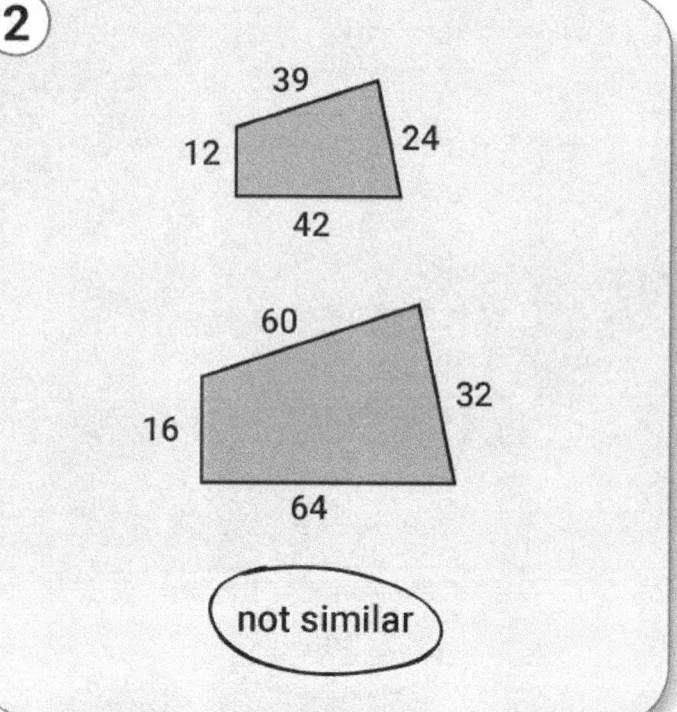

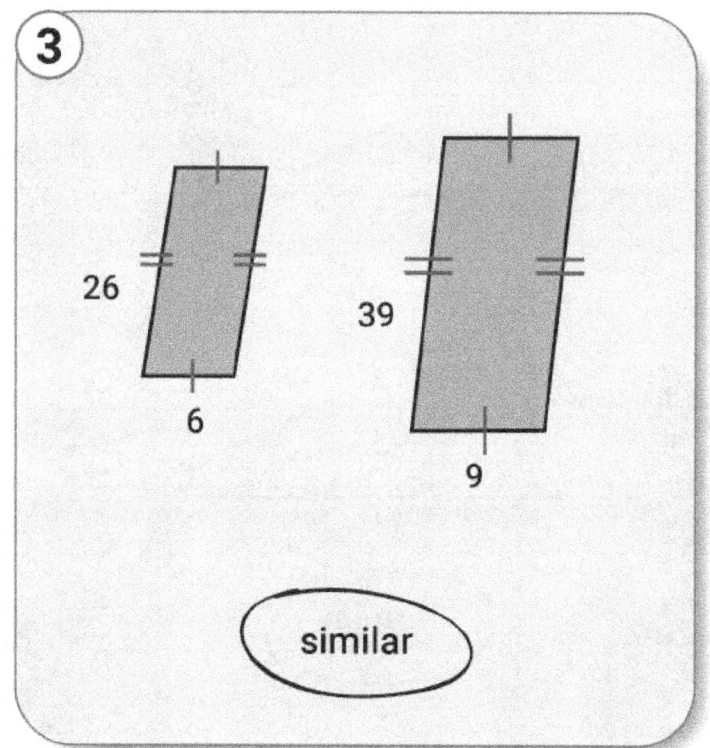

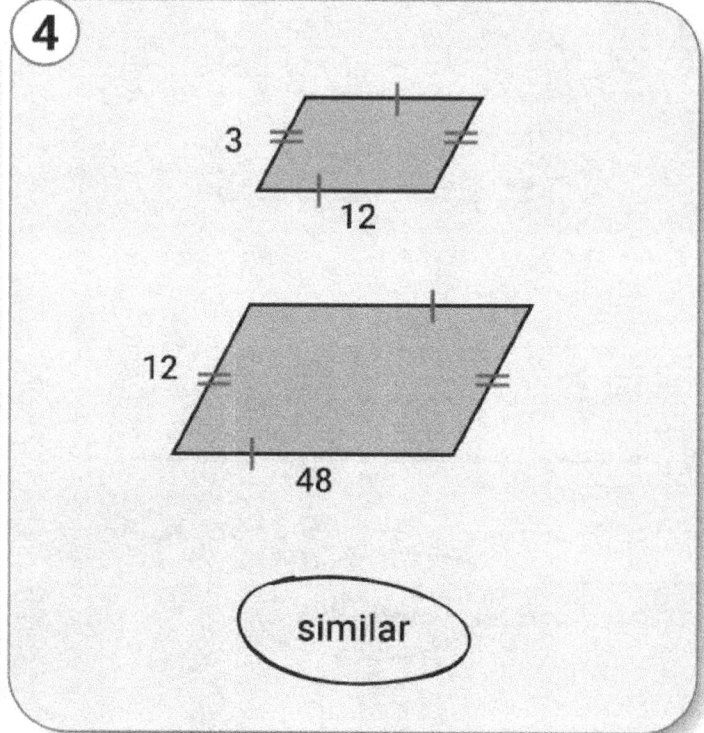

1. not similar

2. not similar

3. similar

4. similar

Find the Scale Factor

1)
2, 12
4, 24

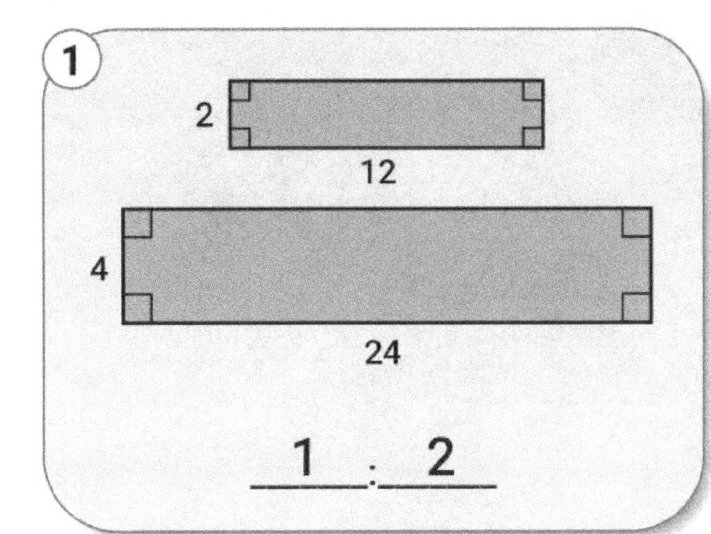

__1__ : __2__

2)
14, 4, 9, 15
42, 12, 27, 45

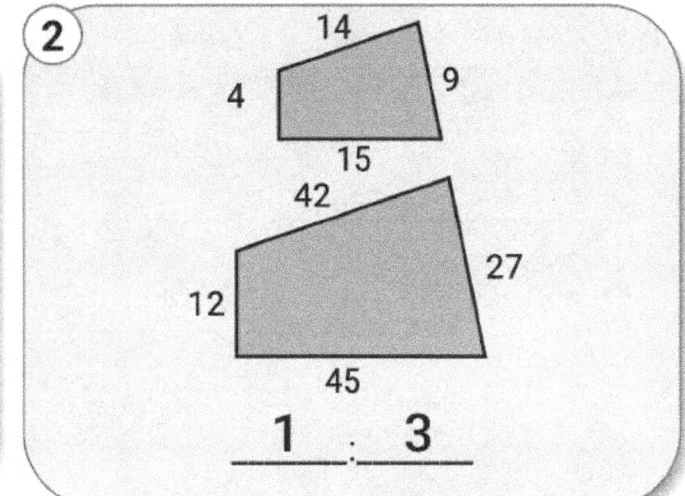

__1__ : __3__

3)
5, 15
30, 90

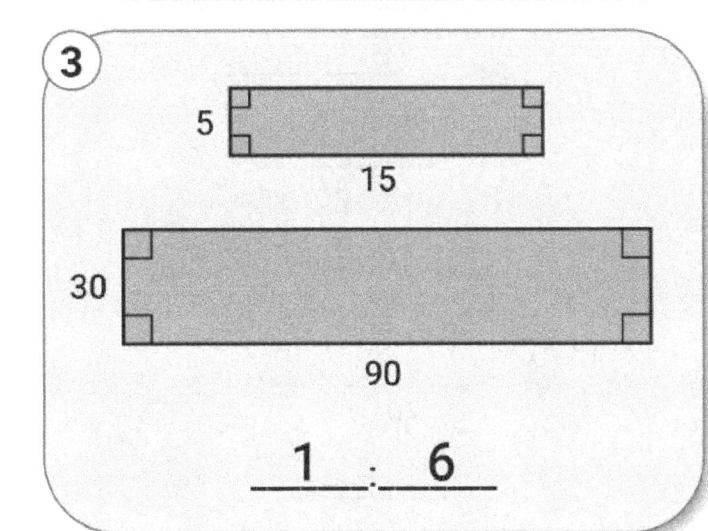

__1__ : __6__

4)
35, 35, 25
42, 42, 30

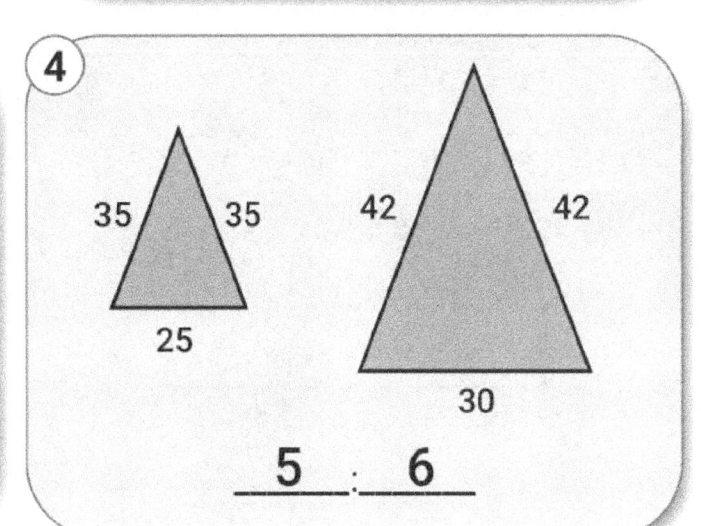

__5__ : __6__

5)
9, 36
15, 60

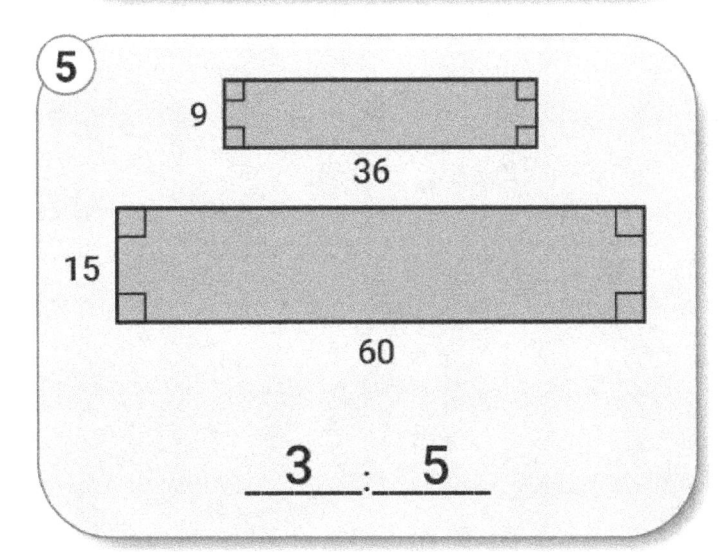

__3__ : __5__

Find the Scale Factor

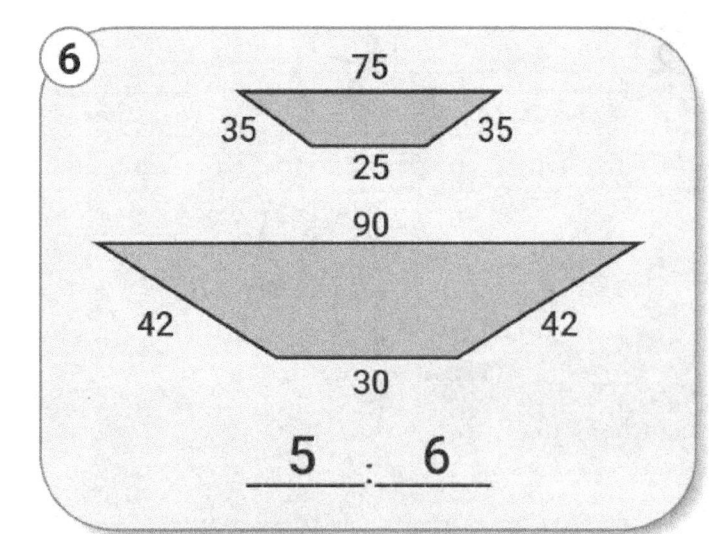

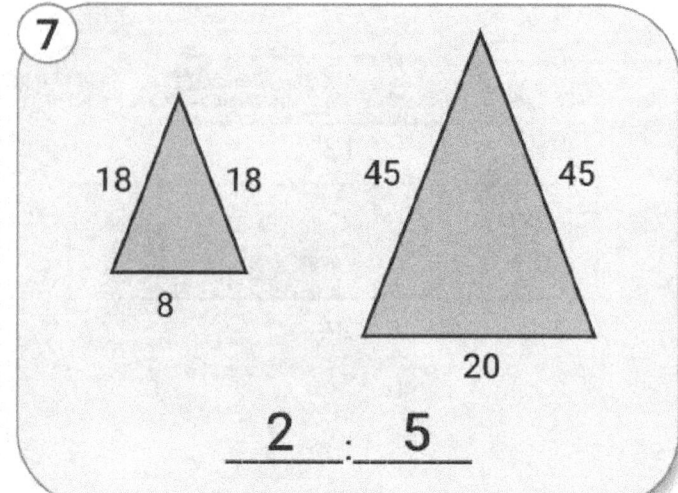

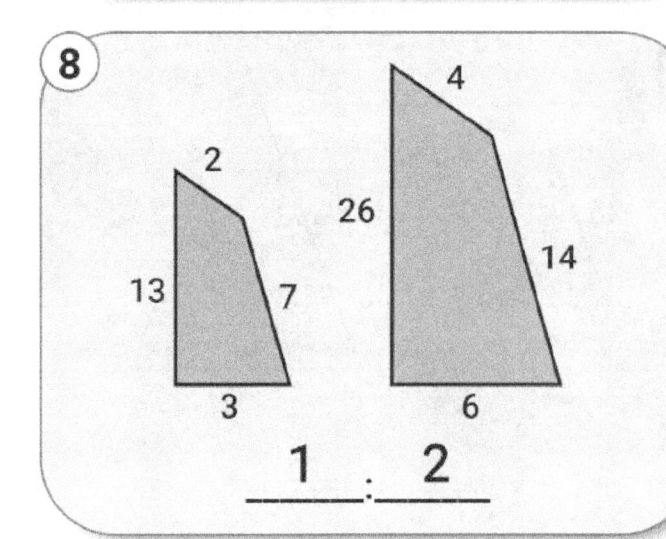

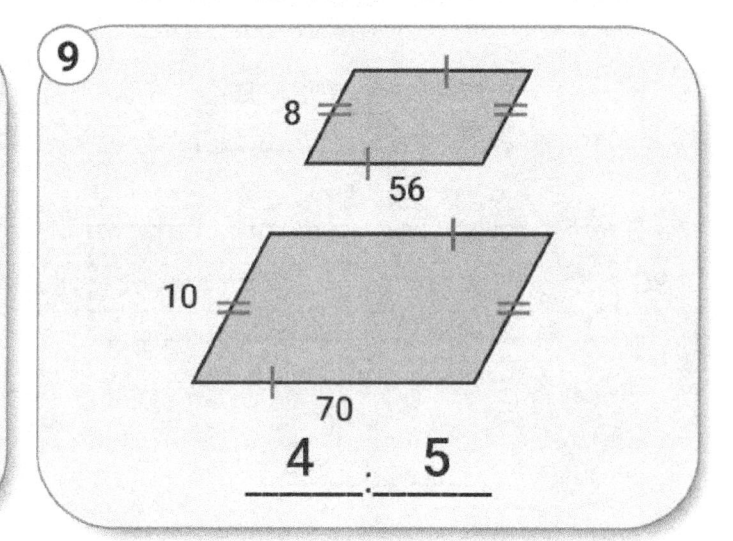

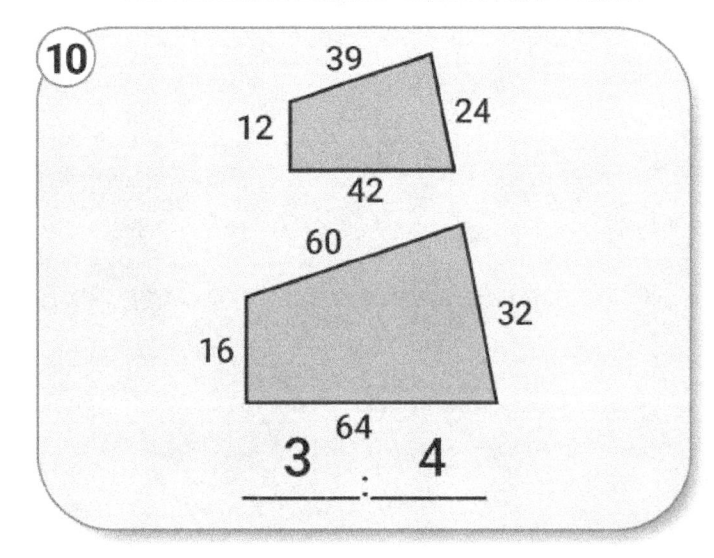

6 5 : 6

7 2 : 5

8 1 : 2

9 4 : 5

10 3 : 4

The Pythagorean Theorem

1) 9 inches (c² = 85, c = 9.2 in.)

2) 14 inches (c² = 200, c = 14.1 in.)

3) 7 inches (c² = 52, c = 7.2 in.)

4) 9 inches (c² = 89, c = 9.4 in.)

5) 14 inches (c² = 208, c = 14.4 in.)

Independent Probability

1. 13/52 = 1/4

2. 2/6 = 1/3

3. 12/52 = 3/13

4. 4/52 = 1/13

5. 3/6 = 1/2

6. 5/6

7. 36

8. 1/2 x 1/2 = 1/4

9. 4/6 = 2/3

10. 3/6 x 2/6 = 6/36 = 1/6

Dependent Probability

1. The probability that the first candy is a lemon drops 6/10 or 3/5. The probability that the second candy is a lemon drop is 5/9. Thus, the probability of choosing two lemon drop is 3/5 x 5/9 = 15/45 = 1/3.

2. 4/9 x 3/8 = 12/72 = 1/6

3. 9/25 x 8/24 x 7/23 = 21/575

4. 1/13 x 4/12 = 1/39

5. 5/13 x 4/12 x 3/11 x 2/10 x 1/9 = 120/154440 = 1/1287

Mean, Median, Mode, and Range

1) Mean: 41, Median: 32, Mode: None, Range: 61

2) Mean: 53, Median: 50, Mode: 69, Range: 62

3) Mean: 71, Median: 73, Mode: None, Range: 59

4) Mean: 61, Median: 65, Mode: None, Range: 52

5) Mean: 71, Median: 75, Mode: None, Range: 44

6) Mean: 39, Median: 40, Mode: 44, Range: 43

7) Mean: 44, Median: 39, Mode: 37, Range: 67

8) Mean: 6, Median: 5.5, Mode: None, Range: 170

9) Mean: 15, Median: 36, Mode: None, Range: 158

10) Mean: 50, Median: 80, Mode: None, Range: 154

Patterns of Association in Bivariate Data

Variable Pair if any, exists for the pair.	Association
Person's age and eyesight ability	No significant association
Hours spent studying and exam grade	Positive correlation
Temperature and amount of clothing worn	Negative correlation
Income and calories consumed	No significant association
Population density and crime rate	Positive association
Amount of soda consumed and number of cavities	Positive association
Car insurance cost and number of years without a traffic violation	Negative association
Age and grade point average	No significant association
Customer satisfaction and number of defective products	Negative association
Shoe size and mile running time	No significant association

CPSIA information can be obtained
at www.ICGtesting.com
Printed in the USA
BVHW010233071120
592780BV00016B/575